THE SECRETARY OF TRANSPORTATION
WASHINGTON, D.C. 20590

April 20, 2010

The Honorable Joseph R. Biden
President of the Senate
Washington, DC 20510

Dear Mr. President:

I am pleased to present the U.S. Department of Transportation's (DOT) report, *Transportation's Role in Reducing U.S. Greenhouse Gas Emissions*. This report, which is submitted in response to the requirements of Section 1101(c) of the Energy Independence and Security Act of 2007, is intended to help inform the debate on surface transportation reauthorization and climate change legislation.

The report examines greenhouse gas (GHG) emission levels and trends from the transportation sector and analyzes the full range of strategies available to reduce these emissions. These strategies include: introducing low-carbon fuels, increasing vehicle fuel economy, improving transportation system efficiency, and reducing carbon-intensive travel activity. While the report does not provide recommendations, it does analyze five categories of policy options for implementing the strategies: an economy-wide price signal, efficiency standards, market incentives, transportation planning and funding programs, and research and development.

The Obama Administration and DOT are committed to tackling the climate change challenge. The DOT recently issued new fuel economy standards as part of the first-ever national greenhouse gas and fuel economy program for cars and light-duty trucks. Further, DOT's livability initiative and sustainable communities partnership with EPA and HUD recognize that multi-modal transportation combined with mixed-use development and smart community planning are key to reducing greenhouse gas emissions.

I look forward to working with Congress on the challenge of transportation and climate change.

Sincerely yours,

Ray LaHood

Enclosure

Acknowledgments

This report to Congress was prepared by the U.S. Department of Transportation (DOT) Center for Climate Change and Environmental Forecasting, supported by a consultant team led by Cambridge Systematics, Inc.

Development of the report was guided by the executive level steering committee and the staff level core team of the DOT Center for Climate Change and Environmental Forecasting. Contributing authors and editors from the Center includes: Julie Abraham, Jan Brecht-Clark, Richard Corley, John Davies, Brigid Decoursey, M.J. Fiocco, Aimee Fisher, Mohan Gupta, Carol Hammel-Smith, Tina Hodges, Michael Johnsen, Mark Johnson, Robert Kafalenos, Linda Lawson, Kelly Leone, April Marchese, Katherine Mattice, Lourdes Maurice, Chris McMahon, Wendy Messenger, Camille Mittelholtz, Alexandra Newcomer, Beth Osborne, Robert Ritter, Arthur Rypinski, Michael Savonis, Tim Schmidt, Helen Serassio, A.J. Singletary, JoAnna Smith, Robert Smith, Alan Strasser, Kathryn Thomson, Robert Tuccillo, Diane Turchetta, and David Valenstein. Tina Hodges of the Federal Transit Administration (FTA) served as the project manager for the Center and as a primary author and editor. M.J. Fiocco of the Research and Innovative Technology Administration (RITA) served as the contracting officer's technical representative.

The Cambridge Systematics contract manager was Joanne Potter. The primary report authors from the consultant team were Christopher Porter, Joanne Potter, and Robert Hyman of Cambridge Systematics; and Rick Baker and Richard Billings of Eastern Research Group, Inc. (ERG). Additional contributing authors included Lance Grenzeback, David Jackson, Gill Hicks, Nathan Higgins, Tracy Clymer, and William Cowart of Cambridge Systematics; Beverly Sauer, Amy Stillings, Sarah Cashman, Ian Todreas, Alan Stanard, Scott Fincher, Michael Sabisch, and Sam Milton of ERG; James Winebrake and James Corbett of Energy and Environmental Research Associates, LLC; Dr. Ian Waitz of the Massachusetts Institute of Technology; Dr. Christopher Frey of North Carolina State University; Cindy Burbank of Parsons Brinckerhoff; and James Shrouds.

Additional staff of the U.S. DOT Administrations as well as the U.S. Department of Energy and the U.S. Environmental Protection Agency also provided input. The team is grateful for the assistance of staff from the Electric Power Research Institute's Transportation and Environment Programs, who provided valuable input into the material on electricity generation and electric vehicle efficiency; and for Tom Reinhart of Southwest Research Institute, who provided valuable input into the material on heavy-duty trucks.

About the Center for Climate Change and Environmental Forecasting

The U.S Department of Transportation (DOT) Center for Climate Change and Environmental Forecasting is the focal point in DOT of technical expertise on transportation and climate change. Through strategic research, policy analysis, partnerships, and outreach, the Center creates comprehensive and multimodal approaches to reduce transportation-related greenhouse gases and to mitigate the effects of global climate change on the transportation network. The Center was formally authorized as the Office of Climate Change and Environment in the Energy Independence and Security Act of 2007.

Table of Contents

List of Tables

List of Figures

Executive Summary

INTRODUCTION

This study evaluates potentially viable strategies to reduce transportation greenhouse gas (GHG) emissions. The study was mandated by the Energy Independence and Security Act (P.L. 110-140, December 2007). The Act directed the U.S. Department of Transportation (DOT), in coordination with the U.S. Environmental Protection Agency (EPA) and consultation with the U.S. Global Change Research Program (USGCRP), to conduct a study of the impact of the Nation's transportation system on climate change and strategies to mitigate the effects of climate change by reducing GHG emissions from transportation. This study also examines the potential impact of these strategies on air quality, petroleum savings, transportation goals, costs, and other factors. Each GHG reduction strategy may have various positive impacts (including co-benefits) or negative impacts on these factors. Potential tradeoffs and interdependencies when reducing GHG emissions will need to be considered in order to develop balanced solutions.

This study does not take a position as to which strategy, or collection of strategies, should be adopted to accomplish the Nation's clean energy and GHG reduction goals. Rather, the study attempts to objectively examine numerous proposed strategies and assess their potential to reduce transportation GHG emissions. The assessments are based on published scientific literature, current policy studies, and best professional estimates. Each strategy is assessed relative to projections of future transportation GHG emissions based on U.S. Energy Information Administration Annual Energy Outlook (AEO) estimates.

The study is presented in two parts: Volumes 1 and 2. **Volume 1: Synthesis Report** provides an overview of the study's findings and discusses policy options that Congress may wish to consider to reduce transportation GHG emissions. **Volume 2: Technical Report** provides the technical details of the assessment.

GREENHOUSE GAS EMISSIONS AND GLOBAL CLIMATE CHANGE[1]

The Intergovernmental Panel on Climate Change (IPCC) estimates that in the absence of additional climate policies to reduce GHG emissions, baseline global GHG emissions from human sources will increase between 25 percent and 90 percent between 2000 and 2030, with CO_2 emissions from energy use growing between 40 and 110 percent over the same period. The IPCC projects that global temperatures will rise between 2°F to 11.5°F by 2100, and global sea level will rise between 7 to 23 inches. More recent estimates that include the effects of polar ice sheet melting suggest a possible 3 to 4 foot sea level rise. According to the Intergovernmental Panel, global GHG emissions must be reduced to 50 to 85 percent below year 2000 levels by 2050 to limit warming to 2.0°C to 2.4°C (3.6°F to 4.3°F). To reach this target, GHG emissions from all sectors must be reduced through a multi-generational effort.

U.S. TRANSPORTATION GREENHOUSE GAS EMISSIONS[2]

The primary greenhouse gases produced by the transportation sector are carbon dioxide (CO_2), methane (CH_4), nitrous oxide (N_2O), and hydrofluorocarbons (HFC).[3] Carbon dioxide, a product of fossil fuel combustion, accounts for 95 percent of transportation GHG emissions in the United States.

Transportation GHG emissions account for 29 percent of total U.S. GHG emissions, and over 5 percent of global GHG emissions.[4] Except otherwise noted, the estimates in this report account for "tailpipe" emissions from burning fossil fuels to power vehicles and do not account for greenhouse gases emitted through other transportation lifecycle processes, such as the manufacture of vehicles, the extraction and refining of fuels, and the construction and maintenance of transportation infrastructure.[5] Including these processes, U.S.

---

[1] Vol. 1, Sec. 2.1. The citation for all figures in this subsection is Vol. 1, Sec 2.1.

[2] Vol. 1, Sec. 2.2.

[3] This report focuses only on emissions of greenhouse gases themselves, rather than emissions of chemicals that interact with other chemicals in the atmosphere to create GHGs.

[4] Vol. 1, Sec 2.2. The citation for all figures in this subsection is Vol. 1, Sec 2.2. Base data is from U.S. EPA (2008). *Inventory of U.S. Greenhouse Gas Emissions and Sinks: 1990 to 2006.*

[5] Life cycle emissions are discussed in Vol 1, Sec. 2.3.

transportation lifecycle greenhouse gases are estimated to account for about 8 percent of global GHG emissions.

Transportation GHG emissions have been growing steadily in recent decades. From 1990 to 2006 alone, transportation GHG emissions increased 27 percent, accounting for almost one-half of the increase in total U.S. GHG emissions for the period.

In 2006, emissions from on-road vehicles accounted for 79 percent of transportation GHG emissions. Emissions from light-duty vehicles, which include passenger cars and light duty trucks (e.g., sport utility vehicles, pickup trucks, and minivans) accounted for 59 percent of emissions. Emissions from freight trucks accounted for 19 percent, and emissions from commercial aircraft (domestic and international) for 12 percent. Emissions from all other modes accounted for less than 10 percent of total emissions.

STRATEGIES TO REDUCE TRANSPORTATION GREENHOUSE GAS EMISSIONS

The study evaluated four groups of strategies to reduce transportation GHG emissions:

- Introduce low-carbon fuels;

- Increase vehicle fuel economy;

- Improve transportation system efficiency; and

- Reduce carbon-intensive travel activity.

The study also evaluated two cross-cutting strategies:

- Align transportation planning and investments to achieve GHG reduction objectives; and

- Price carbon.

Introduce Low-Carbon Fuels[6]

Petroleum-based fuels presently account for 97 percent of U.S. transportation energy use. Low-carbon fuel strategies include the development and introduction of alternative fuels that have lower carbon content and generate fewer transportation GHG emissions. The alternative fuels evaluated in this report include ethanol, biodiesel, natural gas, liquefied petroleum gas, synthetic fuels, hydrogen, and electricity. Alternative fuels strategies have primarily been investigated and quantified for the light-duty vehicle (LDV) sector, although some advances could potentially be applied to other sectors as well.

---

[6] Vol. 2, Sec. 2.

- Renewable fuels such as ethanol and biodiesel offer potential for GHG emission reduction. The GHG emissions benefits of biofuels depend on a variety of factors, including the feedstock, production method, carbon intensity of energy used in production, prior land use, and the evaluation timeframe. Advanced biofuels from cellulosic sources will likely offer much steeper GHG reductions than first generation biofuels, though more research and development is needed, and commercialization has not yet occurred at high volumes. Existing vehicles can operate with low blends of ethanol and biodiesel, but vehicle modifications are needed for higher blends. Adequate distribution of infrastructure is also a key factor. A detailed analysis of renewable fuels is not provided in this report due to rulemaking in this area and readers are directed to http://www.epa.gov/OMS/renewablefuels/ for more information.

- In the long-term (25 years or more), if technical successes in fuel cell development and low-carbon hydrogen production, distribution, and on-board storage can be achieved, hydrogen fuel cell vehicles could reduce per-vehicle GHG emissions by 80 percent or more. Aggressive deployment could reduce total transportation emissions by 18-to-22 percent in 2050 if a 60 percent LDV market penetration could be achieved, which is the optimal end discussed in the literature.[7]

- If significant advances were to occur in battery technology and the use of low-carbon energy sources for electricity generation, electricity (through battery-electric vehicles) could also substantially reduce transportation GHG emissions by 80 percent or more per vehicle in the long term. Aggressive deployment could reduce total transportation emissions by 26-to-30 percent in 2050 if a 56 percent LDV market penetration could be achieved, which is the optimal end discussed in the literature.

Increase Vehicle Fuel Economy[8]

Vehicle and fuel efficiency strategies include developing and bringing to market advanced engine and transmission designs, lighter-weight materials, improved vehicle aerodynamics, and reduced rolling resistance, which would result in lower fuel use and reduced transportation GHG emissions. Many of these technological improvements (such as hybrid-electric powertrains, truck aerodynamic improvements, and more efficient gasoline engines) are well-developed and could be further incorporated into new vehicles in the near future. In the long-term, propulsion systems relying on more efficient power conversion and low- or zero-carbon fuels (such as hydrogen fuel cells or plug-in

---

[7] National Research Council (NRC) (2008). *Transitions to Alternative Transportation Technologies – A Focus on Hydrogen.*

[8] Vol. 2, Sec. 3.

hybrids) may be developed. Many of these strategies have the potential to provide net cost savings as fuel cost savings over the life of the vehicle outweigh the higher cost of the technology. The speed of market penetration of new technologies is limited by the turnover time of the fleet. Passenger cars and light trucks last about 16 years on average before retirement, compared to 20 years or more for trucks, up to 40 years for locomotives and marine vessels, and about 30 years for aircraft.

- Increased fuel economy in light-duty vehicles could reduce total transportation GHG emissions significantly. On a per vehicle basis, compared to a conventional vehicle, GHG reductions are 8-to-30 percent for advanced gasoline vehicles; about 16 percent for diesel vehicles; 26-to-54 percent for hybrid electrics; and 46-to-75 percent for plug-in hybrid electrics.

- Retrofits can be used to expedite improvements. Heavy-duty trucks retrofitted to use aerodynamic fairings, trailer side skirts, low-rolling resistance tires, aluminum wheels, and planar boat tails can reduce per truck GHG emissions by 10-to-15 percent. For new trucks, combined powertrain and resistance reduction technologies are estimated to reduce per vehicle emissions by 10 to 30 percent in 2030.

- Significant fuel economy improvements could also be realized in the rail, and marine sectors—perhaps 20 percent per vehicle for rail and marine and 1.4-2.3% annual improvement for aircraft during 2015-2035—through more efficient engines and resistance reduction technologies. However, total gains are somewhat limited due to the relatively smaller contributions of these sectors.

Improve Transportation System Efficiency[9]

System efficiency strategies reduce the energy use and GHG emissions of travel by optimizing the design, construction, operation, and use of transportation networks.

- Lowering speed limits on national highways would generate moderate and immediate benefits, reducing total transportation GHG emissions by up to 2 percent depending upon enforcement and compliance.

- Strategies such as traffic management and bottleneck relief—including targeted capacity increases at points where demand exceeds capacity—have the potential to modestly reduce GHG emissions by decreasing fuel consumption associated with congestion and stop-and-go traffic (congestion wastes nearly 3 billion gallons of fuel each year[10]). These strategies can also

---

[9] Vol. 2, Sec. 4.

[10] *See* "What Does Congestion Cost Us?" in *2009 Urban Mobility Report*, published by the Texas Transportation Institute.

provide significant cost savings to travelers and businesses. However, the initial GHG benefits of these measures may be partly or fully offset by additional travel resulting from improved travel conditions, known as "induced demand." Because of the uncertainty presently associated with these potentially offsetting effects, the GHG impact of these strategies is not quantified in this report. The DOT is designing research to gain a better understanding of the role of induced demand in offsetting GHG improvements from congestion reduction strategies.

- Direct routing and more efficient takeoff and landing profiles could potentially increase air traffic operational efficiency by 2.5 to 6 percent through 2035. However, these benefits could be offset by induced demand effects which were not quantified for aviation.

Reduce Carbon-Intensive Travel Activity[11]

These strategies would reduce on-road vehicle-miles traveled by reducing the need for travel, increasing vehicle occupancies, and shifting travel to more energy-efficient options that generate fewer GHG emissions. The collective impact of these strategies on total U.S. transportation GHG emissions could range from 5-to-17 percent in 2030, or 6-to-21 percent in 2050.

- Transportation pricing strategies, such as a fee per vehicle-mile of travel (VMT) of about 5 cents per mile, an increase in the motor fuel tax of about $1.00 per gallon, or pay-as-you-drive insurance—if applied widely—could reduce transportation GHG emissions by 3 percent or more within 5-to-10 years. Lower fee or tax levels would result in proportionately lower GHG reductions.

- Significant expansion of urban transit services, in conjunction with land use changes and pedestrian and bicycle improvements, could generate moderate reductions of 2 to 5 percent of transportation GHG by 2030. The benefits would grow over time as urban patterns evolve, increasing to 3-to-10 percent in 2050. These strategies can also increase mobility, lower household transportation costs, strengthen local economies, and provide health benefits by increasing physical activity.

- Studies based on limited European experience suggest that "eco-driving" strategies to teach efficient driving and vehicle maintenance practices could potentially reduce emissions by as much as 1-to-4 percent. However, this would require comprehensive driver training as well as in-vehicle instrumentation. As such, the European findings may not be replicable in the United States.

---

[11]Vol. 2, Sec. 5.

Transportation Planning and Investment[12]

Transportation planning and investment decisions can integrate transportation and land use planning to reduce travel distances, fund low carbon alternatives, and improve the operating efficiency of the multimodal transportation network .

Coordinating transportation and land-use decisions and investments enhances the effectiveness of both, and increases the efficiency of federal transportation spending. In most communities, jobs, homes, and other destinations are located far away from one another, often necessitating a separate car ride for every errand and long delivery routes for goods.   Strategies that support mixed-use development, mixed-income communities, and multiple transportation options can enable travelers to lower trip lengths, reduce trip frequencies, and select more carbon efficient means of travel. These changes in behavior would lower household transportation costs and reduce dependence on foreign oil, while also reducing greenhouse gas emissions. Mixed-use development combined with an increased transit market share may also improve access to jobs and opportunities for those that rely on public transportation.

Planning and investment that increases the share of transportation utilizing low carbon alternatives can reduce GHGs. Examples include public transportation, pedestrian facilities for biking and walking, and lower carbon freight options, including rail or marine.

System efficiency strategies also have potential for GHG reduction and can be instituted through transportation planning processes. These strategies include signal timing, real-time traveler information, more effecive incident management, freeway ramp meeting, and other intelligent transportation systems applications.

There are a range of options for the Federal government to work with State and local governments to incorporate climate change considerations into transportation planning and investment decisions.

Price Carbon[13]

Increasing the cost of carbon economy-wide, through a cap and trade system or carbon tax, provides an economic incentive to consumers and businesses to reduce CO_2 emissions. Policies to price carbon emissions affect all four strategy groups by encouraging use of low carbon fuels and energy efficient vehicles, spurring efficiency improvements in transportation systems, and reducing travel demand.   A cap and trade system consistent with recent proposals could potentially reduce transportation GHG by about 4 percent in 2030, relative to the baseline, and more in future years.

---

[12]Vol. 1, Sec 4.

[13]Vol. 1, Sec. 4.

The intent of pricing carbon is to shift activities to lower carbon alternatives. The availability of alternatives to carbon-intensive travel are crucial to the ability of pricing strategies to reduce carbon emissions in the transportation sector without harming quality of life or the economy. These alternatives include purchasing more fuel efficient vehicles, using lower carbon fuels, taking public transportation or intercity rail, telecommuting, carpooling, and compact development that reduces the need to travel long distances. Without alternatives, consumers are faced with higher costs or reduced quality of life.

FEDERAL POLICY OPTIONS TO ACHIEVE KEY STRATEGIES[14]

Individually and in combination, many of the strategies discussed could significantly reduce transportation greenhouse gases emissions. As Congress considers policy options to pursue, it should be noted that the U.S. Department of Transportation has already committed to pursuing sustainability and livability in transportation programs and making these issues central elements of the surface reauthorization legislation. These elements are critical to achieving a reduction in the GHG emissions of the transportation sector, more transportation choices, and lowering household costs for transportation; while retaining the unique characteristics of our neighborhoods, communities, and regions.

A variety of the strategies discussed in this report are already reflected in DOT's work as it continues to focus on ways to reduce growth in VMT, integrate land use, transportation planning, and investment; and implement system efficiencies necessary to reduce GHG emissions from transportation. In addition, DOT is working on reducing aviation greenhouse gases including developing more efficient aircraft and engine technologies, adopting more energy efficient operational procedures, and advancing the use of renewable fuels.

Building on this work and on the findings of this report, several categories of policy options can be applied to implement the strategies analyzed in this report. Each strategy—vehicle efficiency, low carbon fuels, system efficiency, and reducing carbon intensive travel activity—would require government policies for implementation and to achieve GHG reductions beyond the business-as-usual scenario. This report does not provide recommendations. Instead, it analyzes the potential of each strategy and the policy options for implementing them.

Five broad categories of prospective policy action at the federal level are identified below that could implement the strategies analyzed in this report. The approaches discussed below may be pursued individually or jointly, and in many cases would have synergistic or reinforcing effects when implemented together.

---

[14]Vol. 1, Sec. 5.

Efficiency standards

New standards for fuels and vehicles can achieve significant reductions in carbon emissions from transportation by decreasing the amount of carbon consumed per mile of travel. There is strong evidence that, on average, regulations can achieve fuel consumption and emission reductions while delivering net cost savings to consumers over the life of the vehicle. Equally important, standards would help stimulate research and development. By way of example, the National Highway Transportation Safety Administration (NHTSA) and the United States Environmental Protection Agency (EPA) are working in concert to develop a consistent, harmonized national program that will deliver substantial improvements in fuel economy and reductions in GHG emissions for new cars and light-duty trucks.

Transportation planning and investment allocation policies

Federal transportation planning and investment programs can support integrated transportation and land use planning, provide alternatives to carbon intensive travel, and improve the efficiency of the system—all of which will reduce greenhouse gas emissions. There are three main ways in which the federal government can influence GHG reduction through transportation planning and infrastructure investment: technical assistance, federal transportation planning regulations, and aligning incentives for the tens of billions of dollars of federal transportation investment provided each year.

The U.S. DOT and other federal agencies can provide technical assistance to help transportation agencies conduct GHG inventories and analysis, improve data collection and modeling techniques, and consider GHG emissions in scenario planning, visioning, and integrated transportation and land use planning.

There are a range of options for incorporating climate change considerations. Options range from including GHG emissions as a planning factor, to requiring states and MPOs to develop strategies for reducing transportation GHGs, to establishing mandatory GHG reduction targets. Each option will have differing levels of impact on GHG emissions and require different levels of effort.

Finally, federal transportation funding programs can provide incentives for GHG reduction. Funding incentives could take the form of competitive pools of funding that encourage projects and programs to reduce GHGs. Another option is to align federal funding for transportation infrastructure with performance-based criteria, including climate change objectives that reward effective GHG emission reductions plans and programs.

Market-based incentives

Several market signals specific to the transportation sector could be used to encourage consumers and businesses to more quickly adopt less carbon-intensive vehicles and technologies. By increasing demand for low-carbon technologies, these market signals would spur more rapid private sector research and development. Consideration could be given to continuing and expanding Federal incentives such as those in the Energy Independence and Security Act of 2007.

At the consumer level, rebates and "feebates" could encourage the purchase of high-efficiency and noncarbon-based vehicles. When appropriate, increased motor fuel taxes, variable road pricing, or VMT fees could provide incentives to travelers to reduce trip lengths and shift to less carbon-intensive modes. Tax incentives or low-interest loans for energy-efficient retrofits and new vehicles in heavy-duty, rail, air, and marine sectors, could encourage cross-sector efficiency improvements. Further analysis is needed on options for encouraging fuel efficiency in the rail, marine and aviation sectors, and the potential impacts of these actions.

Research and development

A strong Federal program of interdisciplinary research and technology deployment can advance the effectiveness of the transportation sector in addressing climate change. This research could include both basic and applied research on fuels and vehicles; development of decision support data and tools; research on relationships between climate change and transportation, including risk and adaptation analysis; development of information technologies to support system efficiency; policy research on the interactions among GHG reduction strategies, economic impacts, and institutional issues; and research on equity implications, such as mitigating or avoiding any negative equity impacts from transportation GHG reduction strategies.

Economy-wide price signal

The implementation of carbon pricing — assuming a sufficiently strong price is established — would result in reductions in fuel consumption and an ongoing shift to non-carbon-based fuels and technologies across all sectors. Over the long-term, a cap and trade policy should reinforce technological advances and promote efficiencies in transportation. In order to achieve steep reductions in the transportation sector, complementary policies in addition to a cap and trade system may be required.

CONCLUSION

The ingenuity of transportation planners and engineers has produced a vast network of transportation infrastructure and services to support the mobility and economic vitality of the Nation. However, our historic approach to transportation and land use has created an energy-intensive system dependent on carbon-based fuels and automoibles.

Our national talents and resources must now focus on shaping a transportation system that that serves the Nation's near and long-term goals, including meeting the climate change challenge.

The analysis provided by this report to Congress evaluates the greenhouse gas emission reduction potential of numerous strategies, as well as the co-benefits, costs, and implementation considerations linked to these strategies.

The U.S. Department of Transportation is committed to reducing the impact of the Nation's transportation system on climate change and is already taking action. The Department's livability initiative, along with the Sustainable Communities Partnership with the EPA and HUD, supports low carbon transportation options, such as public transportation, walking and biking. The partnership also promotes mixed-use development that enables residents to easily access goods and services. As shown by this study, all of these actions can reduce greenhouse gas emissions. The Department's high-speed rail initiative will also provide a low carbon travel alternative. Furthermore, in April 2010, the Department and EPA announced the final rulemaking for a national greenhouse gas and fuel economy program for cars and light-duty trucks. The DOT also received new statutory authority under the Energy Independence and Security Act of 2007 to create a fuel efficiency program for medium and heavy duty vehicles and work trucks, which will result in new regulations. In aviation, DOT has put energy and environmental concerns at the core of NextGen—the initiative to modernize the U.S. air traffic system. Likewise, the Maritime Administration is focused on the potential of new technologies to reduce harmful emissions from marine diesel engines through cooperative efforts with the EPA and maritime industry.

Yet there is more to be done. The DOT looks forward to working with Congress on transportation policies that will reduce greenhouse gas emissions, facilitate economic vitality, and enhance our quality of life.

1.0 Introduction

Transportation is a significant contributor to national greenhouse gas emissions, and can be part of the Nation's solution to the climate change challenge. The Energy Independence and Security Act (December 2007) called upon the U.S. Department of Transportation (DOT), in coordination with the U.S. Environmental Protection Agency (EPA) and in consultation with the U.S. Global Change Research Program (USGCRP), to conduct a study of the impact of the Nation's transportation system on climate change and strategies to mitigate the effects by reducing greenhouse gas emissions. The study also considers fuel savings and air pollution reduction from these measures.[15]

This report responds to that directive. *Volume 1: Synthesis Report* provides an overview of transportation's contribution to greenhouse gas emissions (GHG), analyzes the effectiveness of various strategies available to reduce transportation sector GHGs, discusses the role of DOT planning and funding programs for strategic action on climate change, and concludes with five policy options that Congress may wish to consider. Following this introduction:

- **Section 2—Climate Change, Greenhouse Gas Emissions, and Transportation:** Summarizes the effects of transportation emissions on climate change and relative levels of emissions from each transportation mode: cars and trucks, buses, rail, aviation, marine, and pipelines.

- **Section 3—GHG Reduction Strategies:** Discusses the full range of strategies that transportation can employ to directly reduce greenhouse gas emissions from mobile sources across all modes. These strategies are:

  - Introduce low-carbon fuels;

  - Increase vehicle fuel efficiency;

---

[15]P.L. 110-140 states "(c) Transportation System's Impact on Climate Change and Fuel Efficiency – (1) Study. The Office of Climate Change and Environment, in coordination with the Environmental Protection Agency and in consultation with the United States Global Change Research Program, shall conduct a study to examine the impact of the Nation's transportation system on climate change and the fuel efficiency savings and clean air impacts of major transportation projects, to identify solutions to reduce air pollution and transportation-related energy use and mitigate the effects of climate change, and to examine the potential fuel savings that could result from changes in the current transportation system and through the use of intelligent transportation systems that help businesses and consumers to plan their travel and avoid delays, including Web-based real-time transit information systems, congestion information systems, carpool information systems, parking information systems, freight route management systems, and traffic management systems."

- Improve transportation system efficiency; and

- Reduce carbon-intensive travel activity.

The legislative mandate for the report focused on transportation system efficiency and travel activity strategies. As such, the Department made a strong effort to thoroughly cover these areas. The Department also took a holistic approach, including consideration of vehicle technology and alternative fuels strategies. This broad approach better addresses the topic of transportation and climate change, given the important contribution of technology, the interactions between strategies, and the recognition that Federal actions—such as pricing—will often spur both technological and behavioral changes. Even so, due to fuel economy and renewable fuel standard rulemakings, potential transportation greenhouse gas reduction and cost-effectiveness estimates are not presented for some vehicle and fuel technology strategies.

- **Section 4—Cross-cutting Policies:** Discusses policies that involve all four strategies above.

  - *Federal transportation planning requirements and funding mechanisms* can play a role in shaping sustainable transportation programs that provide mobility, while reducing greenhouse gas emissions. These cross-cutting policies influence both system efficiency and travel activity.

  - *Carbon pricing* through a cap and trade system, carbon tax, or a higher motor fuels tax would provide incentives to consumers and businesses to pursue all four strategies above; and

- **Section 5—Policy Options:** Discusses five key Federal policy options that can reduce greenhouse gas emissions from transportation.

Further detail on this analysis is provided in *Volume 2: Technical Report*. Volume 2 contains detailed technical discussions of the four strategy groups that can contribute to reducing the carbon footprint of the transportation sector. All transportation sub-sectors are considered in this report, including on-road vehicles, rail, aviation, and marine. Each set of strategies is evaluated based on a set of factors including magnitude of GHG reduction; timing of impacts, cost, co-benefits (such as fuel savings and air quality) implications for other DOT goals; impacts on infrastructure financing; and feasibility and implementation considerations. The benefits of the strategies in this report are based on limited data and good faith assumptions. Numerical estimates contain substantial uncertainties, as described in the methodology section. Each GHG reduction strategy may have various positive (co-benefits) or negative impacts on other factors as well. Potential tradeoffs and interdependencies when reducing GHG emissions will need to be considered when developing balanced solutions.

Readers interested in the technical basis for the summary materials and policy recommendations contained in *Volume 1* are directed to *Volume 2* for additional background information.

2.0 Climate Change, Greenhouse Gas Emissions, and Transportation

2.1 CLIMATE CHANGE AND GREENHOUSE GASES

Greenhouse gases (GHG) trap heat in the earth's atmosphere. Common greenhouse gases include carbon dioxide (CO_2), methane (CH_4), nitrous oxide (N_2O), ozone, water vapor, and chlorofluorocarbons. Many of these are naturally occuring and necessary to maintain an atmospheric temperature that supports human life.

GHGs are produced by both natural and human activities, and can be removed from the atmosphere through natural processes. However, human-produced GHGs have significantly exceeded natural absorption rates since the industrial revolution. CO_2 is the predominant GHG from human sources, with a majority resulting from the combustion of fossil fuel. Once released, CO_2 and other greenhouse gases take many years to leave the atmosphere. Atmospheric lifetimes are estimated to be 50-200 years for CO_2, 9-15 years for CH_4, and 120 years for N_2O.[16] The combination of long atmospheric lifetimes, increasing GHG output and deforestation have resulted in the increased atmospheric concentration of these gases. Since the beginning of the industrial revolution, atmospheric concentrations of CO_2 have increased by 36 percent, CH_4 concentrations have more than doubled, and N_2O concentrations have risen by approximately 18 percent.[17] Human activities over the past 70 years have also produced synthetic chemicals that are powerful greenhouse gases with atmospheric lifetimes ranging from years to millennia.[18] These substances include hydroflurocarbons (HFCs), chlorofluorocarbons (CFCs) and sulfur hexafluoride (SF_6).

GHG emissions are projected to continue to rise. The Intergovernmental Panel on Climate Change (IPCC) estimates that in the absence of additional climate policies to reduce GHG emissions, baseline global GHG emissions will increase

---

[16] Intergovernmental Panel on Climate Change (IPCC) (1996) Second Assessment Report.

[17] Intergovernmental Panel on Climate Change (IPCC) (2007) Fourth Assessment Report, Climate Change 2007: Synthesis Report. Valencia, Spain.

[18] Intergovernmental Panel on Climate Change (IPCC) (1996) Second Assessment Report.

anywhere from 25 to 90 percent between the years 2000 and 2030, with CO_2 emissions from energy use growing between 40 and 110 percent over the same period.

According to the Intergovernmental Panel, "Warming of the climate system is unequivocal, as is now evident from observations of increases in global average air and ocean temperatures, widespread melting of snow and ice, and rising global average sea level."[19] The panel projects that global temperatures will rise 2-to-11.5°F by 2100, and global sea level will rise 7-to-23 inches (the range of results represents uncertainty in both future anthropogenic emissions and climate modeling). More recent research, including the effects of polar ice sheet melting, suggest that sea levels could rise 3-to-4 feet by the end of this century.[20] The IPCC's report also describes the anticipated consequences of climate change along the range of potential temperature increases, showing severe impacts above 2°C (3.6°F). According to the IPCC, global GHGs must be reduced to 50-to-85 percent below year 2000 levels by 2050 to keep warming to 2.0-to-2.4°C (3.6-to-4.3°F).

Changes in global temperature, and associated changes in weather patterns, have a broad impact on ecosystems, food production, coastlines, human settlements, health, and water supply. In North America, the impact of climate change is felt as drought in areas of the West, because of the reduced mountain snowpack; deteriorating forest health from the increased spread of pests, diseases, and forest fires; changes in agricultural productivity; an increase in the frequency and severity of heat waves, which create adverse health issues for people and animals; and an increase in the risk of flooding in coastal communities.[21] According to USGCRP, widespread climate-related impacts are occurring now and are expected to increase. However, the extent of climate change, and its impacts, depends on the choices made today to mitigate human-caused emissions of GHGs.[22]

Climate change also impacts transportation systems. Rising sea levels and more intense storms can cause increased flooding of coastal transportation facilities such as roads, rail lines, airports, and ports. A DOT study of the central U.S.

---

[19] IPCC Synthesis, 2007 (cited).

[20] USGCRP, 2009 (cited).

[21] Intergovernmental Panel on Climate Change (IPCC) (2007). *Climate Change 2007: Impacts, Adaptation, and Vulnerability. Contribution of Working Group II to the Third Assessment Report of the Intergovernmental Panel on Climate Change* [Parry, Martin L., Canziani, Osvaldo F., Palutikof, Jean P., van der Linden, Paul J., and Hanson, Clair E. (eds.)]. Cambridge University Press, Cambridge, United Kingdom.

[22] U.S. Global Change Research Program (2009). *Global Climate Change Impacts in the United States.* Thomas R. Karl, Jerry M. Melillo, and Thomas C. Peterson, (eds.). Cambridge University Press, p.12.

Gulf Coast area found that 27 percent of major roads in the region are vulnerable to a four-foot sea level rise, and almost one-half the rail miles in the region could be impacted by an 18 foot storm surge.[23] Changes in the frequency and intensity of extreme weather events can disrupt aviation operations. Increases in extreme temperature and precipitation events may also necessitate changes in structural design. In the Arctic where many transportation facilities are built on permafrost foundations, thawing permafrost is already damaging roads and airports. Melting sea ice in Arctic summers may eventually open a Northwest Passage sea lane, changing sea shipping routes across the globe while dramatically altering Arctic ecological systems.[24]

This report to Congress focuses on mitigation strategies to reduce transportation GHGs. Other ongoing DOT projects focus on measures to adapt transportation infrastructure to accommodate the effects of climate change, including a major study of the central Gulf Coast region.

2.2 TRANSPORTATION GREENHOUSE GAS EMISSIONS

GHGs are produced from multiple sectors of the economy, including industrial sources, electric power plants, residences, and agriculture; as well as the different transportation modes. Unlike criteria air pollutants, the main GHGs are global in nature. They do not create toxic "hot spots," but rather are well-mixed in the atmosphere in the long-run. Thus, the impacts of one ton of carbon dioxide emissions are the same no matter where it is emitted, or by what sector of the economy. In that sense, the relative effect of transportation emissions on the global climate can be approximated by their relative magnitude compared to all other global emissions.

The primary GHGs produced by the transportation sector are carbon dioxide, methane, nitrous oxide, and hydrofluorocarbons (HFC).[25] Carbon dioxide, a product of fossil fuel combustion, accounts for 95 percent of transportation GHG

---

[23] U.S. Climate Change Science Program (CCSP) (2008). *Impacts of Climate Change and Variability on Transportation Systems and Infrastructure: Gulf Coast Study, Phase I.* A Report by the U.S. Climate Change Science Program and Subcommittee on Global Change Research [Savonis, M.J., V.R. Burkett, and J.R. Potter (eds.)]. United States Department of Transportation, Washington, D.C..

[24] Transportation Research Board (TRB) (2008). *Potential Impacts of Climate Change on U.S. Transportation.* Transportation Research Board Special Report 290, Committee on Climate Change and U.S. Transportation, Transportation Research Board, Division on Earth and Life Sciences, National Research Council, Washington, D.C.

[25] These gases do not have equal global warming potential (GWP), a measure of relative radiative forcing compared to CO_2. Therefore, unless otherwise noted, figures are presented in CO_2 equivalents, or CO_2e. That is, figures for non-CO_2 GHGs are converted into the amount of CO_2 that would cause the same degree of warming.

emissions in the United States, as illustrated in Figure 2.1. Hydrofluorocarbons, which are used in automobile, truck, and rail air conditioning and refrigeration systems, account for another three percent of U.S. transportation emissions. Nitrous oxide and methane, which are both emitted as byproducts of combustion, account for the remainder of the U.S. transportation GHG emissions inventory.[26]

CO_2, CH_4, N_2O and HFCs are all well-mixed in the atmosphere and long-lived, lasting from years to many decades. While these gases account for a majority of observed global warming effects, human activities produce short-lived and spatially variable emissions that may also have a significant warming effect. Two substances closely associated with the transportation sector are tropospheric ozone and black carbon. Because of their short atmospheric lifetime -- which ranges from weeks to months -- and uncertainties about their global warming potential, tropospheric ozone and black carbon are currently not included in official emissions estimates.

Tropospheric ozone is estimated to have the third-largest increase in radiative forcing since the pre-industrial era, behind CO_2 and CH_4.[27] It is produced when precursors such as nitrogen oxide (NOx), carbon monoxide (CO) and non-methane volatile organic compounds (NMVOCs) react with sunlight in the atmosphere. Motor vehicle exhaust accounts for a majority of U.S. NOx and CO emissions, and is also the largest source of NMVOCs.[28] Black carbon is an aerosol that causes warming by both absorbing solar radiation in the atmosphere and by reducing the reflectivity of snow and ice. The net impact of these two warming effects is estimated to be slightly lower than that of ozone, albeit with a higher degree of uncertainty.[29] Black carbon is emitted from incomplete combustion processes, especially the burning of diesel fuel. On-road sources are estimated to account for about half of U.S. black carbon emissions.[30] Unlike CO2 emissions, ozone precursors and black carbon can be restricted by emissions controls. Both have been significantly reduced by earlier control technologies and are expected to be further reduced by EPA regulations. Given the short

---

[26] This does not include fugitive emissions of CH_4 from natural gas pipelines, which EPA associates with the energy sector rather than the transportation sector in its annual GHG inventory. A discussion of fugitive emissions is provided later in this section.

[27] U.S. EPA, (2009) Inventory of U.S. Greenhouse Gas Emissions and Sinks, 1990-2007.

[28] US EPA (2009)

[29] Intergovernmental Panel on Climate Change (IPCC) (2007) Changes in Atmospheric Constituents and Radiative Forcing.

[30] Unger, N., Shindell, D.T., Wang, J.S., 2009. Climate forcing by the on-road transportation and power generation sectors. Atmos. Environ., 43, 3077-3085.

lifespan of black carbon and ozone precursors, reducing their respective emissions would reduce warming within weeks to months.[31]

Figure 2.1   U.S. Transportation Greenhouse Gas Emissions by Gas, CO_2e 2006

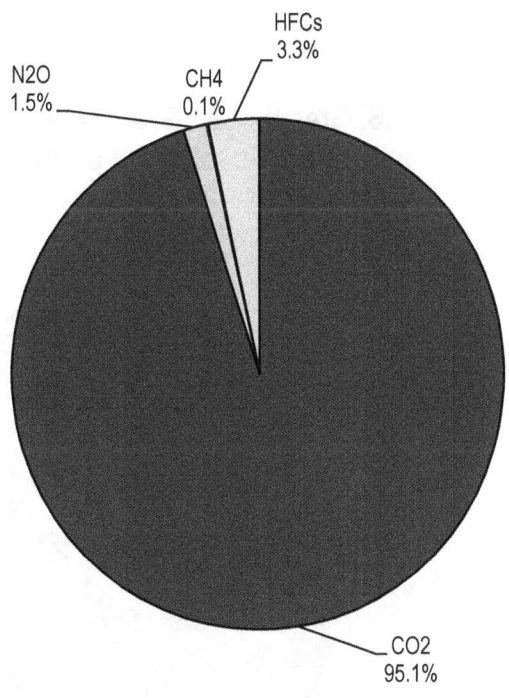

Source: U.S. EPA (2008). *Inventory of U.S. Greenhouse Gas Emissions and Sinks: 1990 to 2006.*

Transportation emissions account for 29 percent of U.S. GHG emissions, and over 5 percent of global GHG emissions.[32] Except when noted, the estimates in this report account for "tailpipe" emissions from burning fossil fuels to power vehicles and not greenhouse gases emitted through other transportation lifecycle processes, such as the manufacture of vehicles, the extraction and refining of

---

[31]Unger, N., Shindell, D.T., Wang, J.S., 2009. Climate forcing by the on-road transportation and power generation sectors. Atmos. Environ., 43, 3077-3085.

Shindell, D., Lamarque, J.F., Unger, N., Koch, D., Faluvegi, G., Bauer, S., Teich, H., 2008. Climate forcing and air quality change due to regional emissions reductions by economic sector. Atmos. Chem. Phys. 8, 7101–7113.

[32] The estimates presented here also include fuels sold in the U.S. to aircraft and ships traveling overseas, also known as international bunker fuels. These estimates are also added to the U.S. total for all sectors. See Table 2.1, U.S. EPA *Inventory of U.S. Greenhouse Gas Emissions and Sinks: 1990 to 2006.*

fuels, and the construction and maintenance of transportation infrastructure. Collectively, emissions from these processes, many of which occur overseas, increase the U.S. transportation share of global GHGs to about 7 to 8 percent.[33] Most of the domestically produced emissions are included in the industry sector shown in Figure 2.2. However, providing a transportation life-cycle estimate provides a broader perspective on the actual impact of transportation on GHGs. A full discussion of transportation life-cycle emissions can be found in Section 2.3.

Figure 2.2   U.S. Greenhouse Gas Emissions by End Use Economic Sector, million metric tons CO$_2$ equivalent
2006

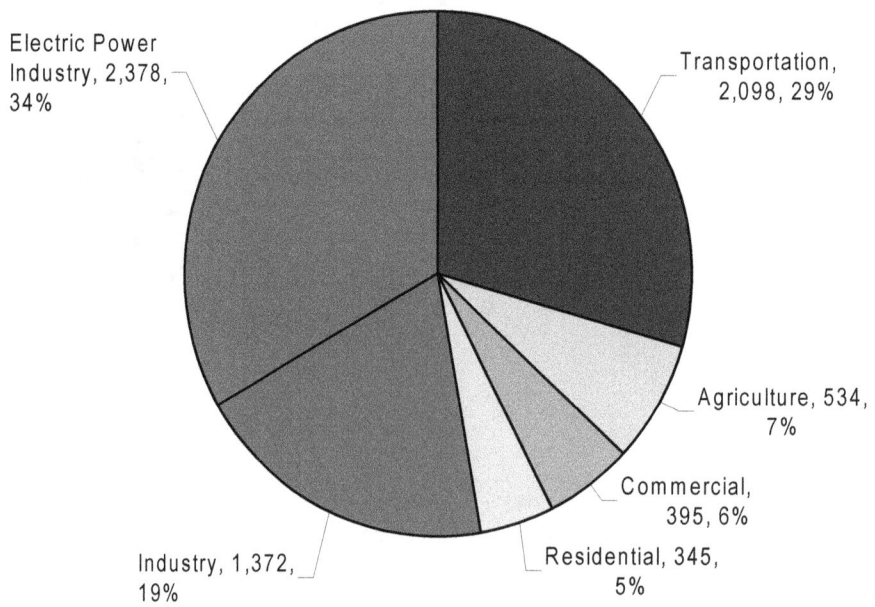

Source: U.S. EPA (2008). *Inventory of U.S. Greenhouse Gas Emissions and Sinks: 1990 to 2006.*

When compared with transportation emissions from all countries in the world, and total world emissions, GHG emissions from the U.S. transportation sector are put into a global context. International Energy Agency (IEA) data for 2006 show that while the U.S. accounts for 5 percent of the world population, it accounts for 21 percent of global CO$_2$ emissions, with the U.S. transportation

---

[33] This is based on the conclusion, as discussed in Section 2.3, that lifecycle emissions may be on the order of 50 percent greater than direct transportation emissions alone.

sector accounting for 33 percent of global transportation CO_2 emissions. Overall, direct emissions from the U.S. transportation sector represent about 7 percent of global CO_2 emissions.[34] GHG emission reduction solutions developed for the U.S. transportation sector could have a significant, direct impact on global GHG emissions. Furthermore, these solutions could also be applied globally to reduce transportation emissions in other countries as well.

As shown in Figure 2.3 and Table 2.1, direct emissions from light-duty vehicles, which include passenger cars and light duty trucks (e.g., sport utility vehicles, pickup trucks, and minivans) accounted for 59 percent of U.S. transportation GHG emissions in 2006. Emissions from freight trucks accounted for 19 percent of emissions. Commercial aircraft (domestic and international) accounted for 12 percent. All other modes accounted for less than 10 percent of total emissions. Overall, on-road vehicles accounted for 79 percent of emissions.

Figure 2.3   U.S. Greenhouse Gas Emissions by Transportation Mode
2006

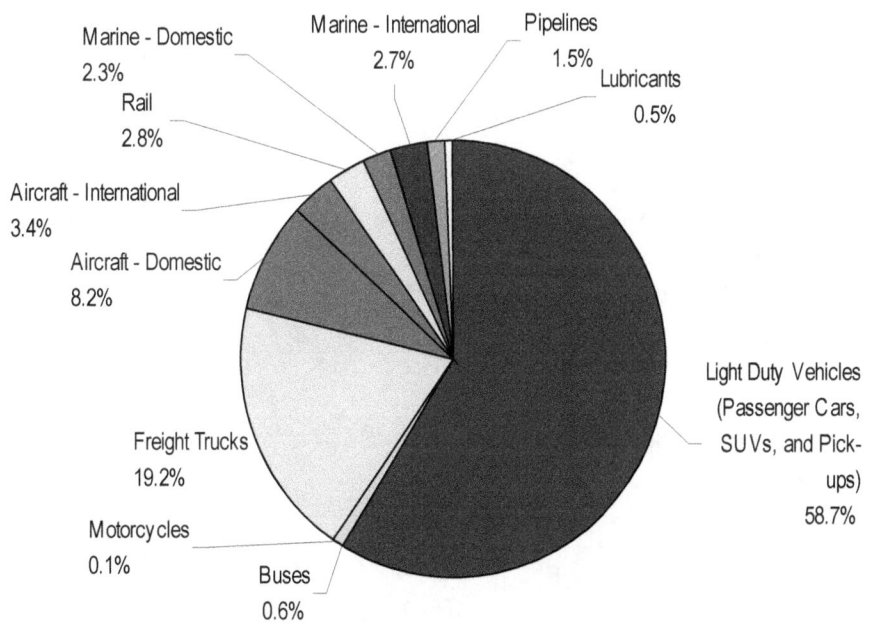

Source: U.S. EPA (2008). *Inventory of U.S. Greenhouse Gas Emissions and Sinks: 1990 to 2006*, pages 3-9, 3-30, 3-31.

---

[34] When including bunkers 2000 EDGAR data show 6.48 percent, 2000 IEA data show 7.89 percent, 2006 IEA data show 6.95 percent. When not including bunkers 2000 EDGAR data show 6.19 percent, 2000 IEA data show 7.27 percent, and 2006 IEA data show 6.46 percent.

Table 2.1 U.S. Transportation Sector Greenhouse Gas Emissions, 2006

Transportation Sources	Million Metric Tons CO$_2$ Equivalent		Change from 1990 to 2006	
	1990	2006	Absolute	Percent
On-Road Vehicles	**1231.9**	**1653.9**	**422.0**	**34.3%**
Light-Duty Vehicles	993.1	1235.0	241.9	24.4%
Passenger Cars	656.9	678.4	21.5	3.3%
Light-Duty Trucks	336.2	556.6	220.4	65.6%
Motorcycles	1.8	1.9	0.1	6.3%
Buses	8.5	12.5	4.0	46.7%
Medium and Heavy-Duty Trucks	228.6	404.6	176.0	77.0%
Aircraft	**228.1**	**244.3**	**16.2**	**7.1%**
Aircraft (Excluding International Bunkers)	***181.9***	***172.4***	***-9.5***	***-5.2%***
Commercial Aircraft— Domestic	138.1	143.6	5.5	4.0%
General Aviation— Domestic	9.5	13.8	4.3	45.5%
Military Aircraft— Domestic	34.3	15.0	-19.3	-56.3%
Aircraft Bunkers	46.2	71.9	25.7	55.6%
Marine	**115.6**	**104.2**	**-11.4**	**-9.9%**
Marine (Excluding Bunkers)	***47.0***	***47.7***	***0.7***	***1.5%***
Recreational Boats	14.2	17.5	3.3	23.1%
Ships—Domestic	32.8	30.2	-2.6	-7.9%
Ships—Bunkers	68.6	56.5	-12.1	-17.7%
Rail	**38.5**	**57.9**	**19.4**	**50.5%**
Pipelines	**36.1**	**32.4**	**-3.7**	**-10.3%**
Lubricants	**11.9**	**9.9**	**-2.0**	**-16.8%**
Total (Including International Bunkers)	**1662.1**	**2102.6**	**440.5**	**26.5%**
Total (Excluding International Bunkers)	**1547.3**	**1974.3**	**427.0**	**27.6%**
Total U.S.—All Sources (Including Int'l Bunkers)	**6263.1**	**7182.5**	**919.4**	**14.7%**
Total U.S.—All Sources (Excluding Int'l Bunkers)	**6148.3**	**7054.2**	**905.9**	**14.7%**

Source: U.S. EPA (2008). *Inventory of U.S. Greenhouse Gas Emissions and Sinks: 1990 to 2006.*

Note: Does not include emissions from ethanol combustion.
Direct emissions only; does not include other fuel, vehicle, or infrastructure lifecycle emissions.

Table 2.1 shows U.S. GHG emissions from transportation sources in 1990 and 2006. U.S. emissions are displayed with and without bunker fuels, which are fuels used for international transport activity by air or water, and are reported based on the location of fuel sales. Fuel sold in the U.S. to ships or aircraft that are bound for international destinations is counted in U.S. bunker fuel totals, but not included in national totals submitted under the United Nations Framework Convention on Climate Change.

GHG emissions from the U.S. transportation sector have been growing steadily — from 1990 to 2006, transportation GHG emissions increased 27 percent. The growth in U.S. transportation GHG emissions accounted for almost one-half (47 percent) of the increase in total U.S. GHG emissions for the period. Emission trends vary by transportation mode. Medium and heavy-duty truck GHG emissions increased 77 percent from 1990 to 2006, while light duty vehicles increased 24 percent; and aircraft 7 percent. On-road vehicles accounted for 96 percent of the increase in transportation emissions during that period; 55 percent from light-duty vehicles, 40 percent from medium and heavy-duty trucks, and one percent from other modes.

Light-Duty Vehicles

Between 1990 to 2006, an increase in vehicle-miles traveled (VMT) and a stagnation of fuel economy across the U.S. vehicle fleet, caused light-duty vehicle GHG emissions to grow by 24 percent.

VMT increased 39.4 percent between 1990 and 2006, as shown in Figure 2.4, which is over twice the U.S. population growth rate during that period.

Average fuel economy among new vehicles sold showed a very slight decline from 1990 to 2004, and a very slight increase thereafter, as shown in Figure 2.5. The decline in new vehicle fuel economy prior to 2004 reflected the increasing market share of light duty trucks, which grew from about one-fifth of new vehicle sales in the 1970s to slightly over one-half by 2004, as shown in Figure 2.6.

Trends in transportation GHGs can generally be seen as a race between fuel economy and VMT. If VMT growth outpaces improvements in fuel economy, emissions will grow. If fuel economy improvements outpace VMT growth, emissions will decline.

Recent trends indicate that light duty vehicle emissions are leveling off as VMT growth slows and fuel economy improves. Growth in passenger vehicle VMT slowed from an annual rate of 2.6 percent from 1990 to 2004 to an average annual rate of 0.6 percent from 2004 to 2007. In 2008, VMT on all streets and roads in the United States decreased for the first time since 1980, likely due to a combination of high fuel prices and a weakening economy. In addition, average new vehicle fuel economy improved from 2005 to 2007 as the market share of passenger cars increased compared to light-duty trucks; also a response to higher fuel prices and a weakening economy. As discussed in Section 2.4, light duty vehicle GHGs are projected to almost plateau as anticipated VMT growth modestly outpaces new fuel economy and low-carbon fuel standards.

Figure 2.4   Vehicle Miles Traveled by Light Duty Vehicles
1975 to 2008

U.S. Vehicle Miles Traveled (in Millions)

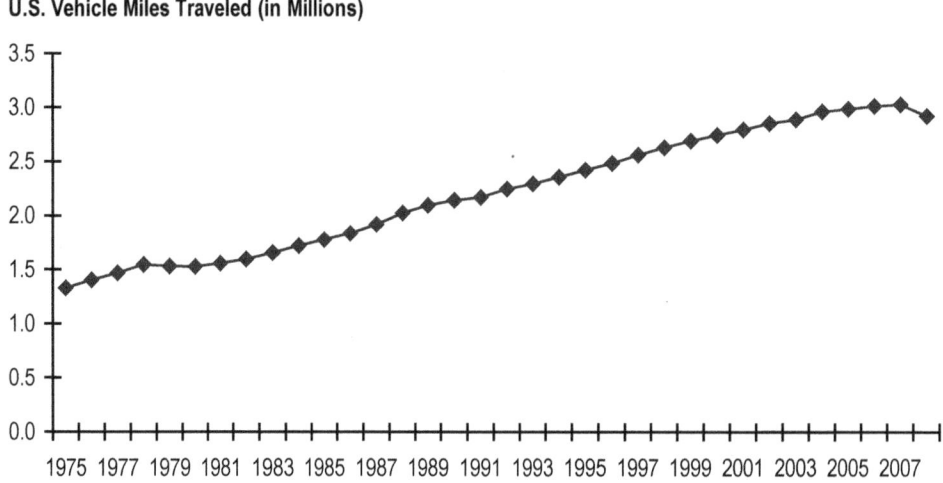

Source: Bureau of Transportation Statistics. National Transportation Statistics.

Figure 2.5   Sales-Weighted Fuel Economy of Light Duty Vehicles

Fuel Efficiency (Miles per Gallon)

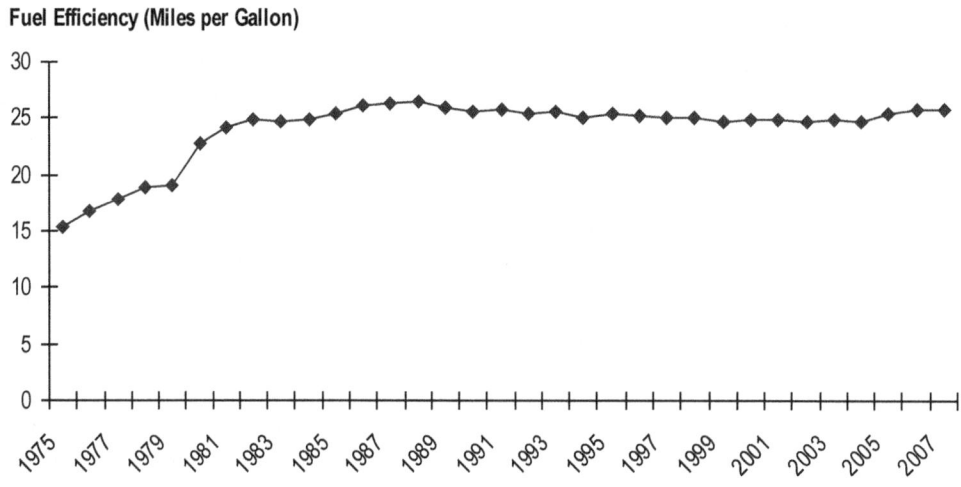

Source: U.S. EPA (2007). *Light-Duty Automotive Technology and Fuel Economy Trends: 1975 Through 2007.*

Figure 2.6   Sales of New Passenger Cars and Light Duty Trucks

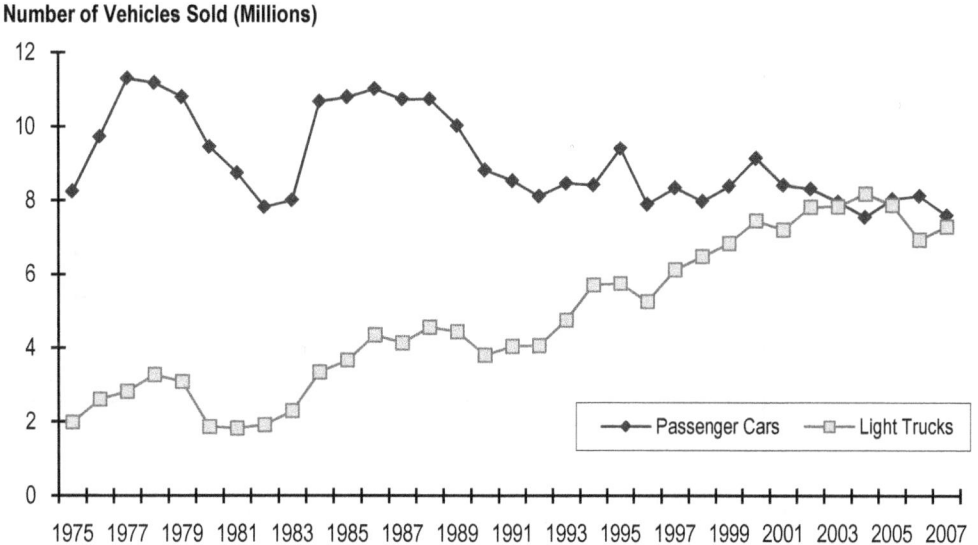

Source: U.S. EPA (2007). *Light-Duty Automotive Technology and Fuel Economy Trends: 1975 Through 2007.*

Medium and Heavy Duty Vehicles

Since 1990, GHG emissions from medium and heavy-duty trucks have increased 77 percent, growing at three times the rate of emissions from light-duty vehicles. This is the product of decreasing fuel efficiency—as measured per ton-mile carried—and steadily increasing demand for freight trucking. Between 1990 and 2005, CO_2 emissions per ton-mile carried increased almost 13 percent, while actual ton-miles carried increased 58 percent.[35] These changes were driven by an expansion of freight trucking after economic deregulation of the trucking industry in the 1980s; widespread adoption of just-in-time manufacturing and retailing practices by business shippers and receivers, increasing highway congestion; and structural changes in the economy that produce higher-value, lower-weight, and more time-sensitive shipments better served by trucking. GHG emissions from freight trucks have increased at a greater rate than all other freight sources, as shown in Figure 2.7.

---

[35] U.S. EPA (2008). *Inventory of U.S. Greenhouse Gas Emissions and Sinks: 1990 to 2006.*

Figure 2.7   GHG Emissions from U.S. Freight Sources

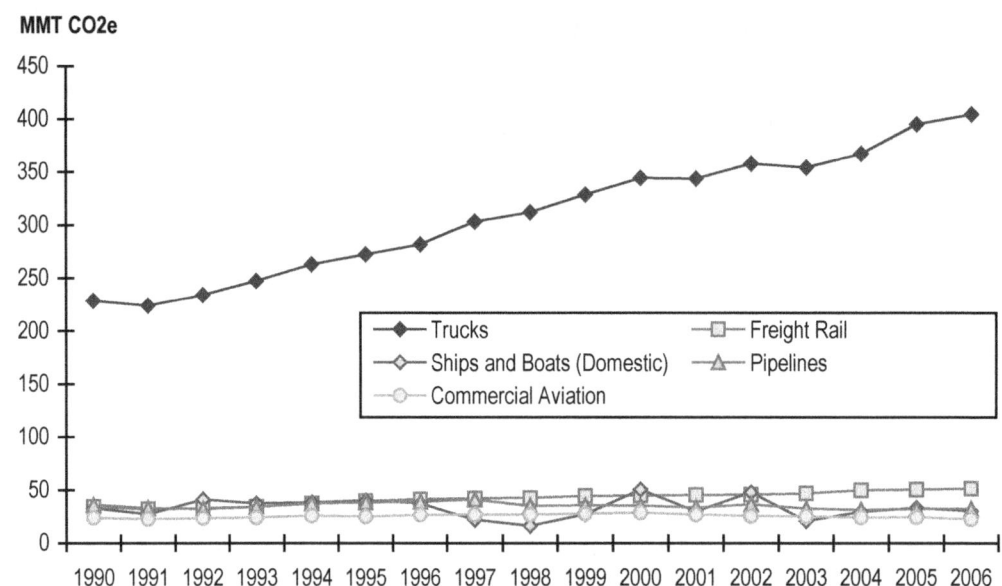

Source: U.S. EPA (2008). *Inventory of U.S. Greenhouse Gas Emissions and Sinks: 1990 to 2006.*

MMT CO2e = million metric tons carbon dioxide equivalent

Aircraft

Although airline passenger miles traveled increased 69 percent between 1990 and 2006 (Figure 2.8), airline GHG emissions increased only 4 percent over the same period.   This is primarily because passenger loads increased substantially, to nearly 80 percent.   Additionally, the energy efficiency of new engines increased by about one percent annually and the energy efficiency of the fleet as a whole improved, due to the accelerated retirement of older aircraft following the terrorist attacks of September 2001.   Eighty-six percent of aircraft emissions are allocated to passenger travel and 14 percent to air cargo.[36]

Because their emissions take place in the upper atmosphere, aircraft have unique effects on climate change beyond the direct effect of the greenhouse gases emitted.   The two primary complexities are due to the generation of nitrogen oxides (NO_x) and water vapor by jet engine combustion.   In the upper atmosphere, NO_x has two opposing effects: it leads to the production of ozone,

---

[36]Since individual aircraft carry both passengers and air cargo (in the belly) simultaneously, emissions are allocated in EPA's emissions inventory using data on the weight of freight shipped and total number of enplaned passengers. (U.S. EPA, 2008, cited).

but at the same time it increases the rate of destruction of methane (CH₄).[37] These effects are quite substantial: ozone production is estimated to be almost as significant as CO_2 emissions in terms of warming potential from aircraft emissions.[38] In addition, the injection of water vapor into the very dry and very cold upper atmosphere may lead to contrail formation, which can have a warming effect by trapping infrared radiation from the ground. It also may lead to increased cirrus cloud formation, which may lead to net warming. Not all of these dynamics are well understood, or well quantified, leading to greater uncertainty when estimating the impact of air travel on global warming.[39]

Figure 2.8 Trends in Passenger Activity and Fuel Efficiency for Aircraft

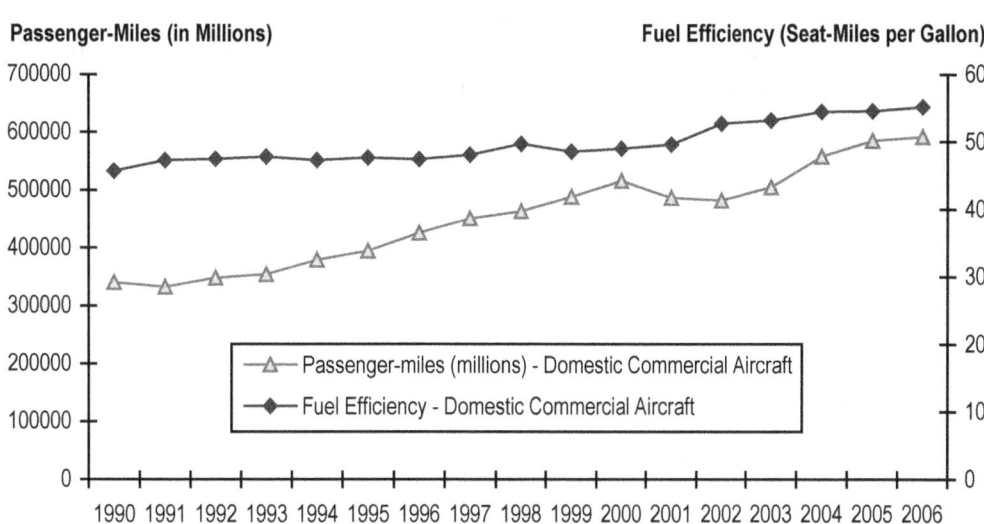

Source: Bureau of Transportation Statistics. National Transportation Statistics.

[37] U.S. EPA (2006). *Greenhouse Gas Emissions from the U.S. Transportation Sector, 1990 to 2003.*

[38] Brasseur, G.P. (ed.) (2008). *A Report on the Way Forward Based on the Review of Research Gaps and Priorities.* Environmental Working Group of the U.S. NextGen Joint Planning and Development Office, Aviation Climate Change Research Initiative, Federal Aviation Administration, Washington D.C.

[39] Brasseur, G.P. (ed.), 2008 (cited).

Rail

GHG emissions from rail primarily originate from the combustion of diesel fuel by locomotives, but 8 percent are also attributed to electrically powered trains.[40] Freight sources of rail emissions include: line-haul trains, which travel long distances on intercity routes; and switchyard locomotives, which move around rail yards to assemble rail cars into trains. Passenger sources of rail emissions include urban transit, commuter and inter-city rail. As shown in Figure 2.9a, GHG emissions from freight rail have steadily increased from 1990 to 2006, while emissions from passenger rail have increased slightly over the same period. Increasing freight rail activity, shown in Figure 2.9b, has led to increased freight rail emissions. However, simultaneous increases in fuel efficiency (also shown in Figure 2.9b) have counteracted this trend to slow the growth of rail GHG. Results of a Federal Railroad Administration (FRA) study indicate that railroads now handle 50 percent more ton-miles of freight than in 1990—using 21.5 percent less fuel per ton-mile.[41]

---

[40] The EPA inventory reports 4.9 million metric tons CO_2e from electricity for the total U.S. transportation sector in 2006. The electricity currently used in transportation is almost entirely for electrically powered trains.

[41] U.S. Department of Transportation, Federal Railroad Administration. *Comparative Evaluation of Rail and Truck Fuel Efficiency on Competitive Corridors.* November 2009.

Figure 2.9 Rail Trends: a) Greenhouse Gas Emissions, and b) Revenue Freight Ton-Miles and Fuel Efficiency
1990 to 2006

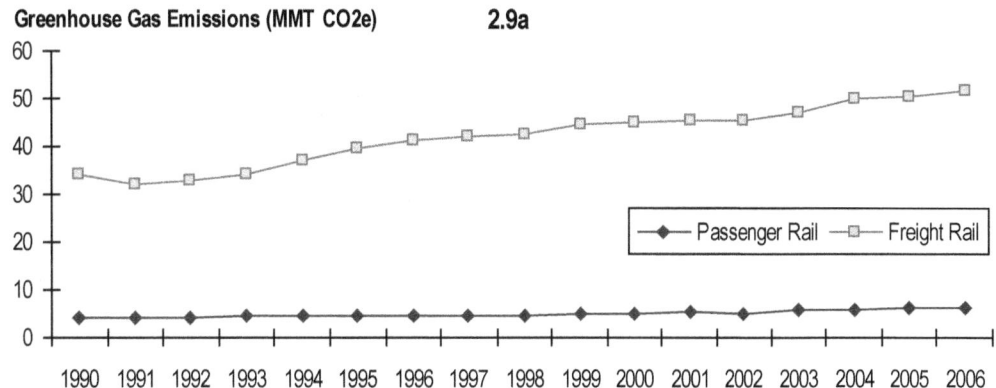

Source: U.S. EPA (2008). *Inventory of Greenhouse Gas Emissions and Sinks: 1990 to 2006.*

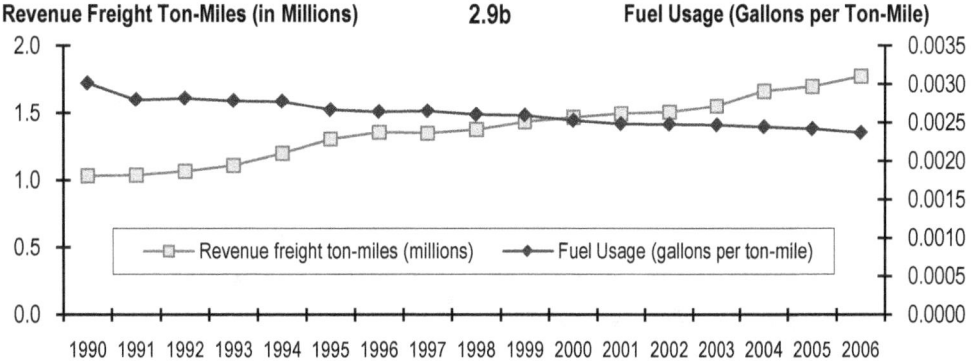

Sources: Bureau of Transportation Statistics. National Transportation Statistics (Freight Activity and Fuel Efficiency).

Marine

Figure 2.10a depicts greenhouse gas emissions from the broad range of marine activities: recreational craft, inland waterway, marine coastal, and international shipping. Greenhouse gas emissions from marine sources appear rather volatile in some years in Figure 2.10a. This fluctuation is most likely due to issues with data collection and interpretation, including the challenge of separating the domestic and international components of fuel consumption estimates. These issues may also reflect the nature of ship refueling strategies, which take advantage of price differences among countries to buy fuel at the least expensive port. As shown in Figure 2.10b, domestic waterborne freight tonnage decreased from 1990 to 2006, while international waterborne imports and exports increased significantly as the Nation's international trade grew. Because of the complex nature of some routing, such as multiple stops to exchange cargo and routing through the Panama Canal, a clear relationship between tonnage and GHGs emitted is difficult to discern.

Figure 2.10 Marine Trends: a) Greenhouse Gas Emissions, and b) Ton-Miles of Freight
1990 to 2006

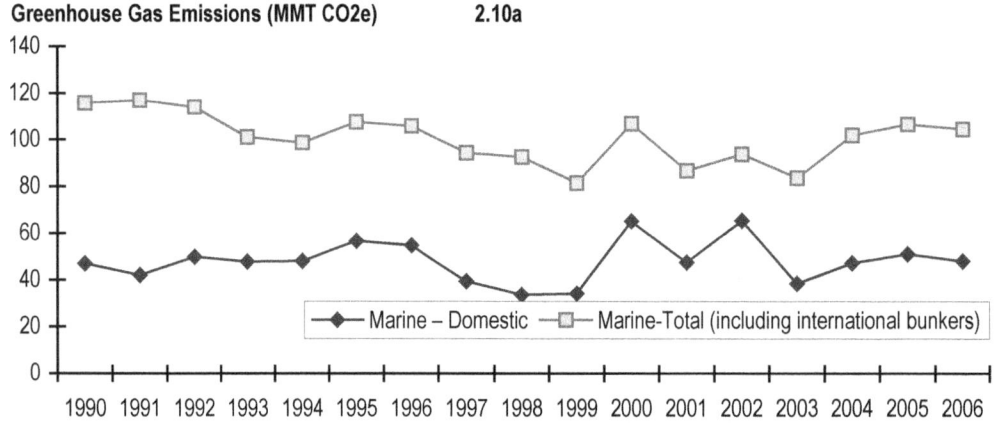

Source: U.S. EPA (2008). *Inventory of U.S. Greenhouse Gas Emissions and Sinks: 1990 to 2006.*

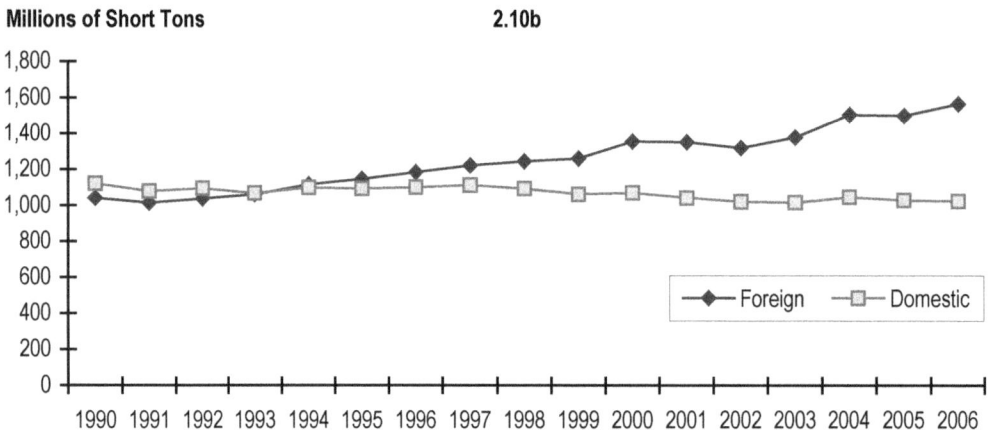

Source: Bureau of Transportation Statistics. National Transportation Statistics.

Pipelines

The Pipeline and Hazardous Materials Safety Administration (PHMSA) is the primary regulatory agency responsible for the construction, safe operation and oversight of our Nation's natural gas and hazardous liquid pipelines.

Pipelines are classified as part of the transportation system because they are used to transport large amounts of natural gas and petroleum products in the United States.   Pipelines are typically powered by pumps, motors, engines, and compressors that run on either natural gas or electricity. GHGs are produced by pipelines from either the combustion processes for these fuels, or leaks—referred to as *fugitive emissions*—from natural gas pipelines.

From 1990 to 2006, GHG emissions associated with powering pipelines have decreased by 10.3 percent—despite a 5.6 percent increase in ton-miles transported by pipeline. However, pipeline GHG estimates only reflect CO_2 emissions from the burning of natural gas to power pipelines and do not include electricity, which is also a major power source for pipelines.[42] Decreasing emissions could represent a shift to electricity to power pipelines, increases in pumping efficiency, or some combination of the two; specific data are not available to provide further detail.

During this same time period, GHG fugitive emissions have lowered by 18.4 percent. This is likely due to improvements in management practices and technologies that help prevent leakages from pipelines and compressor stations.[43]

Transportation GHGs by Freight and Passenger Travel

Another way to understand transportation GHG emissions is to examine the split between passenger and freight transport. In 2006, passenger transportation generated 73 percent of transportation GHG emissions, while freight transportation accounted for 27 percent. At the same time freight transportation GHG emissions have grown more quickly than passenger emissions, and accounted for more than 40 percent of the increase in transportation GHG emissions over 1990 to 2006. The sources of passenger transportation GHG emissions include: passenger cars, light trucks, buses, and motorcycles; most aircraft emissions; and a small portion of rail and marine emissions. The sources of freight transportation GHG emissions are: medium- and heavy-duty trucks; pipelines; a large majority of rail and marine operations; and a small portion of air travel.

Passenger transportation GHG emissions can be viewed in terms of GHG emission per passenger-mile traveled (PMT), which takes into account the different number of passengers carried by each mode. The rates for each of the passenger modes are charted in Figure 2.11. Buses, which include transit, intercity and school buses, produce the least GHG emissions per PMT. They are

---

[42] U.S. EPA, 2008 (cited).

[43] Methane (CH_4) is released from natural gas pipelines as fugitive emissions due to leaks during the transmission, storage, and distribution process. These fugitive emissions are typically accounted for in the energy sector rather than the transportation sector and are not shown in Table 2.1. Their inclusion would increase pipeline GHG emissions from 32.4 mmt CO_2e to 95.4 mmt CO_2e, increasing pipelines' share of U.S. transportation GHGs from 1.5 to 4.4 percent. Source: U.S. EPA, 2008 (cited).

followed by motorcycles, passenger rail (which includes transit, commuter, and intercity rail), commercial aviation, passenger cars, and light duty trucks.

Buses have the lowest emissions per PMT because of their high occupancy rate– an average of about 21 people per bus, when including all types of bus service.[44] Transit buses have a lower occupancy rate—about 9 to 10 people per bus averaged across the U.S.[45] However, transit buses only account for 15 percent of all bus passenger-miles traveled.[46] Intercity passenger rail averages about 20 passengers per car, while rail transit averages 23, and commuter rail averages 31.[47] Light-duty vehicles average 1.6 persons per vehicle.[48] Commercial airliners are very energy intensive per vehicle-mile traveled, but have slightly lower GHG emissions per passenger-mile traveled than light-duty vehicle, due to the high number of passengers per plane. Aircraft and inter-city rail may involve additional auto or transit travel from the airport/station to the final destination; the emissions from this leg of the trip are not included here.

---

[44] U.S. Department of Transportation, Bureau of Transportation Statistics. National Transportation Statistics, http://www.bts.gov/publications/ national_transportation_statistics/. Tables 1-32 and 1-37.

[45] U.S. Department of Transportation, Federal Transit Administration. 2007 National Transit Database.

[46] Davis, S.C., S.W. Diegel, and R.G. Boundy (2008). *Transportation Energy Data Book*. Oak Ridge National Laboratory, Oak Ridge TN (Table 2.12); U.S. Department of Transportation, Bureau of Transportation Statistics. National Transportation Statistics, http://www.bts.gov/publications/national_transportation_statistics/. Table 1-37.

[47] Davis, S.C., S.W. Diegel, and R.G. Boundy, 2008 (cited).

[48] U.S. Department of Transportation, Bureau of Transportation Statistics. 2001 National Household Transportation Survey.

Figure 2.11 GHG Emissions per Passenger-Mile Traveled (PMT) by Passenger Transportation Mode
2006

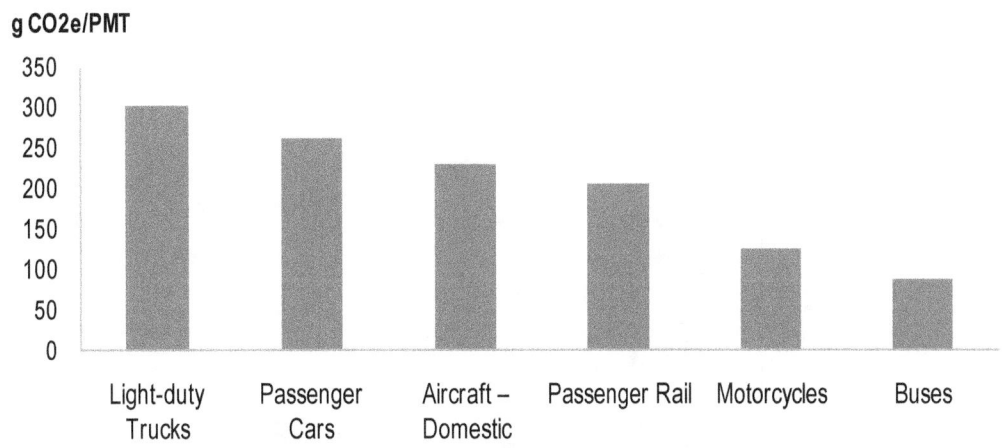

Source: U.S. EPA, *Inventory of U.S. Greenhouse Gas Emissions and Sinks: 1990 to 2006*; Bureau of Transportation Statistics. Bureau of Transportation Statistics, National Transportation Statistics; and U.S. Department of Energy, *Transportation Energy Data Book*.

Freight modes have very different GHG profiles, as shown in Figure 2.12. On a ton-mile basis, freight rail is the lowest emission mode, followed by pipelines and marine transport. These three freight modes specialize in carrying bulk or containerized goods in large quantities, at relatively slow speeds, and achieve significant economies of scale. Trucking generates significantly higher GHG emissions per ton-mile, reflecting the energy inefficiencies of relatively small vehicles traveling at higher speeds, as well as the lighter weight of its cargo. Aircraft, which primarily carry high-value, time-sensitive cargo, have by far the highest GHG emissions per ton of freight.[49] However, because of the different mix of traffic that the modes carry, a head-to-head comparison between modes—based purely on tonnage—may not present a complete picture. A more appropriate comparison would be to consider the energy consumption of the different modes moving similar traffic within specific corridors. For instance, a FRA study compared rail and truck fuel efficiency by focusing on corridor-specific competitive services that each mode provides. Overall, the study found that rail achieved 1.4 to 9 times more ton-miles per gallon than competing truckload service. An update to this study finds that rail-fuel improvements have outpaced truck-fuel improvements over the study period.[50]

---

[49] Including freight carried in the belly of passenger aircraft.

[50] U.S. Department of Transportation, Federal Railroad Administration, 2009 (cited).

Figure 2.12 GHG Emissions per Freight Ton-Mile by Freight Transportation Mode
2006

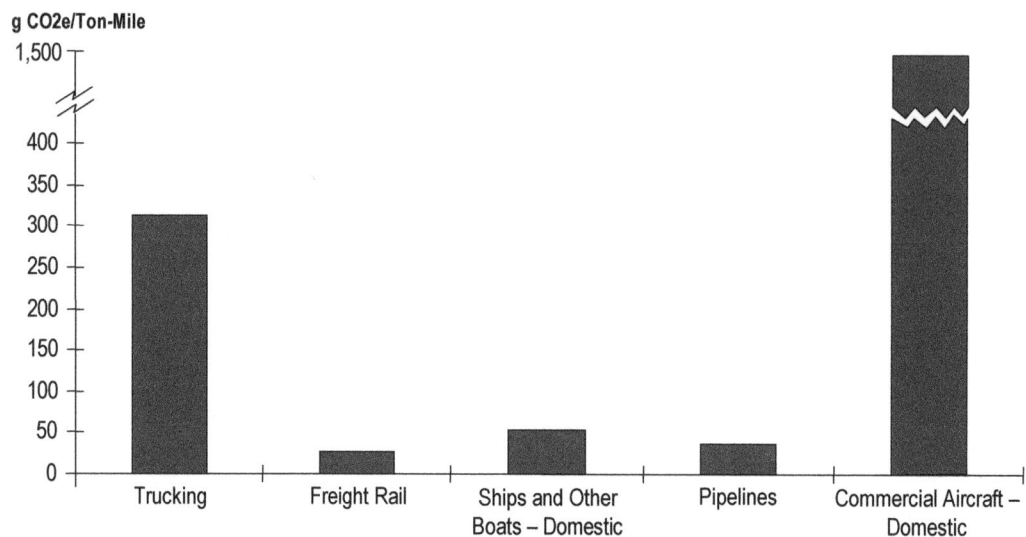

Source: U.S. EPA (2008). *Inventory of U.S. Greenhouse Gas Emissions and Sinks: 1990 to 2006*; and Bureau of Transportation Statistics. National Transportation Statistics.

Transportation, Housing, and Environmental Policy Coordination and Impact on Vehicle Miles Traveled (VMT)

Transportation and land use are interdependent. Decisions on the locations and densities of housing, retail, offices, and commercial properties impact travel patterns to these destinations. Similarly, the geographic placement of roads, public transportation, airports, and rail lines influences where homes and businesses are built. Areas of lower density tend to have higher levels of automobile use per capita.[51] Lack of coordination in location decisions has resulted in more frequent and longer trips, and thus higher GHG emissions. Over the past several decades, housing densities have decreased and the amount of developed land in the country has grown faster than population.[52] In many communities, development has largely been automobile-oriented, such that a car or truck is used by residents for the majority of their travel.

---

[51] See for instance, Peter Newman and Jeffrey Kenworthy (2006) "Urban Design to Reduce Automobile Dependence", Opolis: An International Journal of Suburban and Metropolitan Studies: Vol. 2: No. 1, Article 3. http://repositories.cdlib.org/cssd/opolis/vol2/iss1/art3

[52] National Academy of Sciences, *Transportation Research Board Special Report 298: Driving and the Built Environment*, 2009.

Coordinated planning allows communities to understand the synergies and advantages to planning investments in transportation, housing, and other community amenities together. Coordinated planning processes involve a range of decision-makers, including local land use planning officials, private investors, developers, State departments of transportation, Federal agencies, air quality planning and air quality officials, and community groups. Coordinated planning informs the ultimate decisions that are made, and can affect the long-term impact of transportation on climate change. However, it is important to note that these decisions are made at the local level, many factors are considered, and the Federal government may have limited influence on some of these decisions. Regardless, the Federal government can do more to coordinate Federal housing, transportation and environmental programs and policies. Furthermore, the Federal government can offer technical assistance to local governments to enhance their capacity for more environmentally sustainable investments. Such Federal activities could guide local investment decisions and reduce GHG emissions by enabling more carbon-efficient choices.

Refrigerant GHGs from Mobile Air Conditioners and Refrigerated Transport

In 2006, HFC emissions from mobile air conditioners and refrigerated transport vehicles/containers accounted for 69.5 mmt CO_2e, or 3.5 percent of total transportation GHG emissions. It should be noted that these emissions are included in the estimates for the modes discussed above, but are presented here separately to highlight the special characteristics of the gases and the unique factors that influence the levels of their release.

HFCs were introduced in the early 1990s as a new refrigerant for mobile air conditioners and refrigerated freight transport units to replace chlorofluorocarbons (CFC) and hydrochlorofluorocarbons (HCFC). CFC and HCFC were banned under the Montreal Protocol due to their ability to deplete the ozone layer. All of these refrigerants are very potent greenhouse gases with a high global warming potential (GWP). GWP is a measure used to convert all greenhouse gases into equivalent units based on their ability to trap radiation.[53] HFC-134a, the most commonly used HFC refrigerant today, has a GWP of 1,300, indicating that 1 kilogram of HFC-134a has the same warming potential over a 100-year period as 1 kilogram of CO_2.

HFCs are released into the atmosphere through leaks in mobile air conditioners or refrigerated transport units during operation. Leaks also occur while

---

[53] The GWP of a greenhouse gas is defined as the ratio of the time-integrated radiative forcing from the release of 1 kilogram (kg) of the gas (or other substance) relative to that of 1 kg of CO_2. The GWP of CO_2 is always defined as 1, because it is the reference gas that all others are compared to. GWPs provide a mechanism for converting all greenhouse gas emissions to an "equivalent" amount of CO_2.

servicing these units, or during their retirement; at which time the HFCs can be recycled. Efforts to prevent these leaks, such as strengthening the requirement and training of technicians to use recovery equipment, and not vent refrigerant during equipment service, have reduced emissions. Investigating new refrigerants with lower GWP values could decrease the global warming effect of future releases of current refrigerants.

Figure 2.13 shows the steady increase in HFC emissions from 1990 to 2006 associated with the gradual introduction of HFCs to replace CFCs and HCFCs as common refrigerants. HFC emissions plateau after 2005, possibly due to better leak prevention and completion of the HFC phase-in.

Figure 2.13 HFC Emissions from Mobile Air Conditioners and Refrigerated Transport

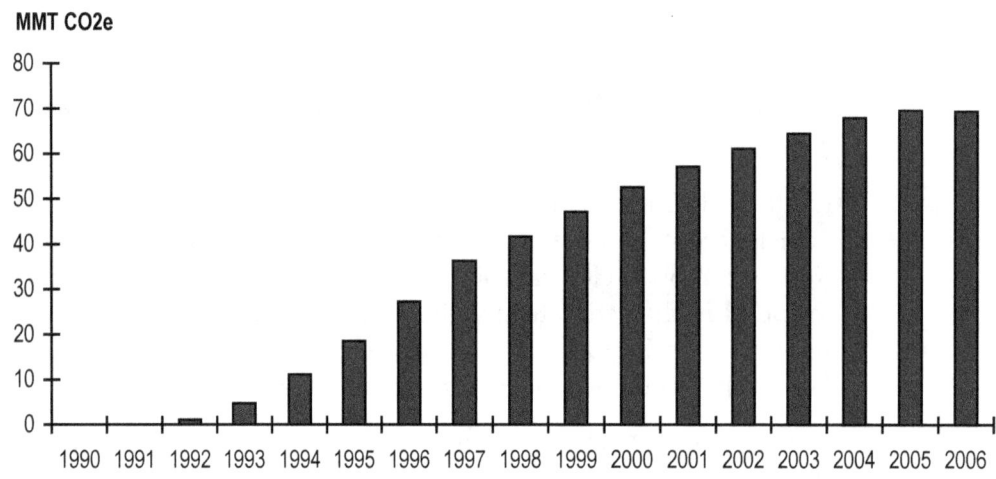

Source: U.S. EPA (2008). *Inventory of U.S. Greenhouse Gas Emissions and Sinks: 1990 to 2006.*

2.3 LIFE-CYCLE TRANSPORTATION GREENHOUSE GAS EMISSIONS

A primary source of transportation greenhouse gases is the combustion of fuel or other energy sources to power vehicles, also known as "tailpipe" emissions. However, transportation depends on an array of other processes that also produce additional GHG emissions. These include the production and distribution of fuel, the manufacture of vehicles, and the construction and maintenance of transportation infrastructure. These supporting processes — known as the fuel, vehicle manufacture, and infrastructure cycles — generally are not included in U.S. transportation sector GHG estimates. Many of these processes are included in U.S. industrial sector estimates, and some occur overseas, and are therefore excluded from estimates of U.S. transportation sector GHG emissions. However, these processes are important elements of the transportation life cycle, increasing GHG emissions by up to 50 percent more than operating emissions alone.

GHGs from these three supporting processes appear to be of comparable magnitude, with fuel cycle emissions likely having the largest contribution. Fuel cycle processes include the extraction, shipment, refining and distribution of fuel, and GHGs from these activities vary by fuel type.[54] Gasoline fuel cycle processes are the most GHG intensive of any conventional transportation fuel, with fuel cycle processes producing GHGs that are roughly 24 to 31 percent beyond the combustion emissions of the fuel itself. Diesel fuel cycle emissions are roughly 15 to 25 percent beyond direct diesel combustion emissions, while jet fuel is 17 to 24 percent beyond combustion emissions.[55]

Vehicle manufacture cycle emissions include raw material production, vehicle construction and shipment. GREET and LEM provide estimates of GHGs from these processes for on-road vehicles; additional estimates are provided by Chester (2008). With these estimates expressed relative to combustion emissions, the manufacture-cycle GHGs represent an additional 14 to 19 percent beyond gasoline combustion emissions; manufacturing of freight trucks is 6 to 17 percent beyond combustion diesel combustion emissions; and aircraft up to 6 percent.

As shown in Figure 2.14, the EPA estimates that greenhouse gas emissions for light-duty vehicles are 38 to 50 percent higher than operating emissions alone when fuel cycle and vehicle cycle emissions are also included. For diesel-powered freight trucks, emissions are 21 to 45 percent greater when including fuel cycle and vehicle cycle emissions.

---

[54] Estimates have been developed for Argonne National Laboratory's GREET model, as well as for the Life-Cycle Emissions Model (LEM) developed by Mark Delucci at the University of California at Irvine.

[55] U.S. Environmental Protection Agency (2006). *Greenhouse Gas Emissions from the U.S. Transportation Sector: 1990-2003.*

Figure 2.14 Direct GHG Emissions Plus Fuel and Vehicle Cycle GHG Emissions

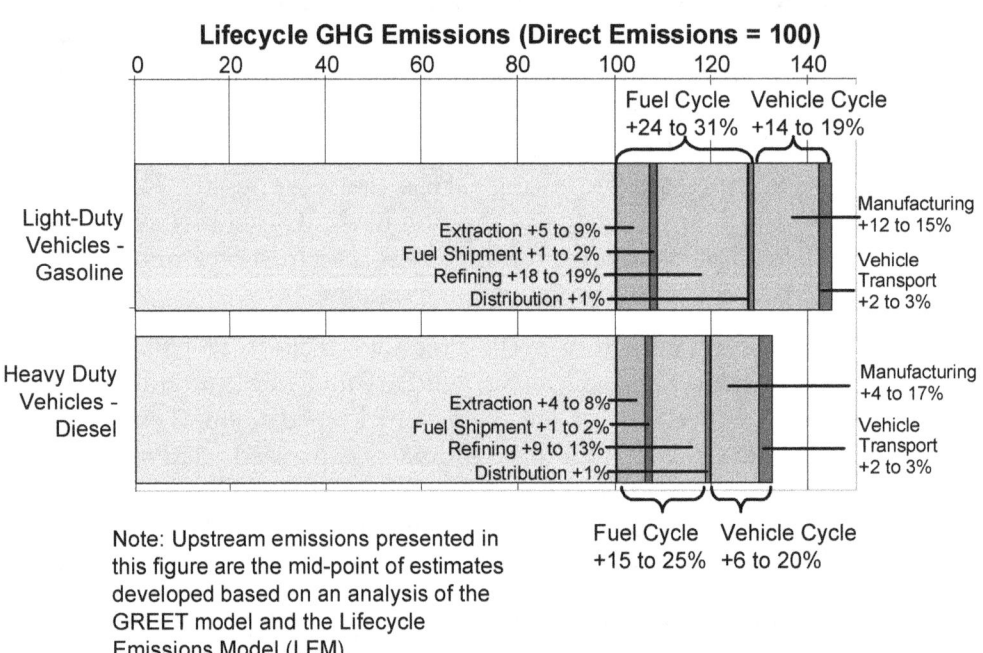

Note: Upstream emissions presented in this figure are the mid-point of estimates developed based on an analysis of the GREET model and the Lifecycle Emissions Model (LEM).

Source: U.S. EPA (2006). *Greenhouse Gas Emissions from the U.S. Transportation Sector, 1990-2003* (EPA 420-R-06-003).

Including infrastructure cycle emissions in estimates would further increase these figures. However, there is limited evidence on vehicle infrastructure cycle emissions, although research in this area has accelerated. The only published estimates incorporating infrastructure cycle emissions, as well as fuel and vehicle cycle emissions, are provided by Chester, as shown in Table 2.2 (for selected transportation modes). These results suggest that all together, fuel, vehicle and infrastructure cycle emissions increase emissions by one-half beyond operating emissions alone for light-duty vehicles and buses; double the emissions for rail transit; and increase aircraft emissions around a quarter.[56] For light-duty vehicles, the contribution of infrastructure construction, operations and maintenance is of roughly the same magnitude as each of the contributions of vehicle manufacturing and fuel production. For rail-transit modes, the relative contribution of infrastructure varies depending upon various factors, such as the amount of tunneling and the elevated right of way used in the system.[57]

---

[56] Chester, Mikhail Vin (2008). *Life-cycle Environmental Inventory of Passenger Transportation Modes in the United States.* Institute of Transportation Studies, University of California, Berkeley.

[57] See Chester (2008), Figures 4 and 11.

Table 2.2   Life-Cycle GHG Estimates for Various Transportation Modes

Vehicle Type	Operational emissions, g CO2e/ PMT	Fuel, Vehicle, and Infrastructure Cycle emissions, g CO2e/PMT	Total emissions, g CO2e/ PMT	Percent Increase over Operational Emissions Alone
On-Road Vehicles				
Sedan	230	150	380	65%
SUV	270	180	450	67%
Pickup	420	200	620	48%
Bus (Off-Peak Times)	470	210	680	45%
Bus (Peak Times)	59	26	85	44%
Rail Transit Systems				
Bay Area Rapid Transit (BART)	64	76	140	119%
Caltrain	74	86	160	116%
San Francisco Municipal Transit Agency (Muni)	69	101	170	146%
Massachusetts Bay Transit Authority (MBTA)—Green Line	120	110	230	92%
Aircraft				
Embraer 145	230	60	290	26%
Boeing 737	170	40	210	24%
Boeing 747	150	50	200	33%

Source: Chester, Mikhail Vin (2008). *Life-Cycle Environmental Inventory of Passenger Transportation Modes in the United States.* Institute of Transportation Studies, University of California, Berkeley.

Notes: Calculations are based on average occupancies. The researchers analyzed particular rail systems as the rail life cycle emissions vary greatly by system.

2.4 PROJECTED GROWTH OF TRANSPORTATION GREENHOUSE GAS EMISSIONS

Based on the Energy Information Administration's (EIA) Annual Energy Outlook (AEO), projections show little growth in GHG emissions from transportation in the coming decades—with total GHG emissions growing only 0.7 percent between 2007 and 2030, as shown in Figure 2.15.[58] CO_2 emissions alone from transportation are expected to grow 1.8 percent, slower than the 3.5 percent growth projected for the economy as a whole.[59]

According to these projections, the modes show very different rates of growth in emissions, as shown in Table 2.3. Despite a 42 percent increase in VMT over the period, light-duty vehicle GHG emissions are projected to decline nearly 12 percent, in response to expected increases in fuel economy from corporate average fuel economy (CAFE) regulations, advanced technologies, and alternative fuels.[60] Freight trucks, on the other hand, show a projected 20 percent increase in emissions, even though freight truck VMT grows at a similar rate to light-duty vehicles. Domestic aviation also shows significant projected growth, with emissions climbing 27 percent. As a result, domestic aviation's share of total transportation emissions is expected to grow from 9 percent to more than 11 percent over this period. Although the share of emissions from light-duty vehicles is projected to decrease, they would still account for nearly one-half of transportation CO_2 emissions.

These projections are subject to a number of uncertainties: economic growth, population growth, future fuel prices, and expected changes in the future mix of vehicles and fuels. The AEO projections are particularly sensitive to the assumed rate of growth in VMT, because on-road vehicles account for more than three-quarters of transportation GHG emissions. Higher or lower VMT projections will significantly change the projections of total GHG emissions from

---

[58] Based on EIA AEO 2009 April Update, Reference case. CO_2 numbers were taken directly from AEO for all modes. CH_4, N_2O, and HFCs were calculated using scaling factors developed from reported shares for each gas from the *U.S. Inventory of Greenhouse Gas Emissions and Sinks 1990-2006* (U.S. EPA, 2008), as well as trends in CH_4 and N_2O emissions. HFC emissions were assumed to remain a constant share of modal emissions.

[59] Energy Information Administration. Annual Energy Outlook 2009 April Update, as cited.

[60] In the AEO 2009 reference case, ethanol GHG emissions are assumed to be net zero in the transportation sector (direct emissions are offset by the growing of the feedstock). The emissions of ethanol production and harvesting are included in the industrial sector.

transportation. This AEO projection assumes an average annual growth in VMT of 1.5 percent.

The AEO projections take into account existing government legislation and regulations, but do not consider additional government policies, such as subsequent increases in fuel economy standards. The projections shown here are based on the AEO reference case from the April 2009 update, which includes the anticipated effects of the American Recovery and Reinvestment Act (ARRA), and the fuel economy and renewable standards included in the Energy Independence and Security Act (EISA) of 2007.

With the new standard of 35.5 mpg for new vehicles by 2016, light-duty GHG emissions would be approximately 4 percent lower in 2020, and 3 percent lower in 2030, than the projections provided in this report.

Figure 2.15 Historic and Projected Transportation GHG Emissions (mmt CO$_2$e)
1990 to 2030

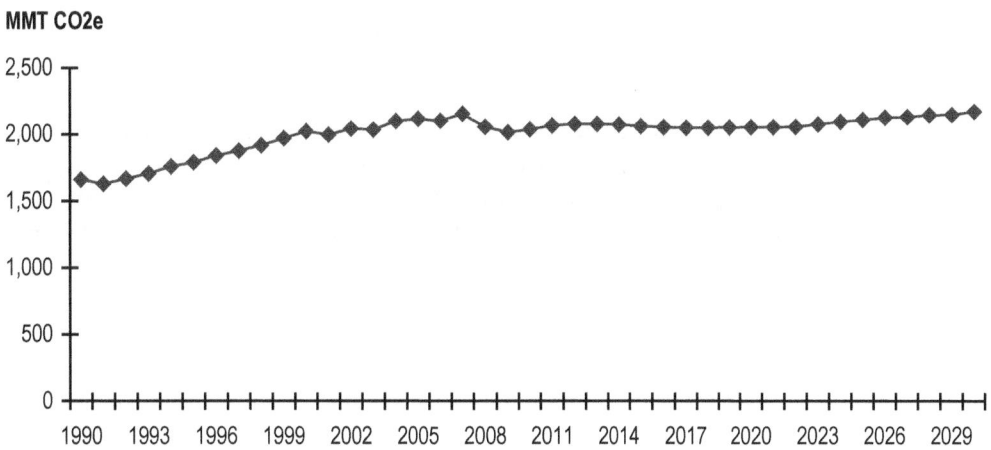

Source: Historical emissions (1990 to 2006) from U.S. EPA (2008): *U.S. Inventory of Greenhouse Gas Emissions and Sinks 1990 to 2006.* Projected emissions (2007 to 2030) from Cambridge Systematics analysis of Energy Information Administration, 2009: Annual Energy Outlook 2009 Updated April Release.

Table 2.3   GHG Emission Projections By Mode (mmt CO₂e)
2007 to 2030

	2007	2030	Percent Change 2007-2030	2007 Share by Mode	2030 Share by Mode
Light-Duty Vehicles	1,221.4	1,080.9	-11.5%	56.7%	49.8%
Commercial Light Trucks	43.4	41.6	-4.3%	2.0%	1.9%
Bus Transportation	20.2	20.6	2.0%	0.9%	0.9%
Freight Trucks	374.9	449.7	20.0%	17.4%	20.7%
Rail, Passenger	6.6	8.2	24.7%	0.3%	0.4%
Rail, Freight	48.8	55.4	13.5%	2.3%	2.6%
Shipping, Domestic	28.3	32.7	15.7%	1.3%	1.5%
Shipping, International	78.0	79.9	2.5%	3.6%	3.7%
Recreational Boats	19.7	21.2	7.8%	0.9%	1.0%
Air	194.1	246.6	27.1%	9.0%	11.4%
Military Use	50.3	55.2	9.8%	2.3%	2.5%
Lubricants	5.2	5.6	7.5%	0.2%	0.3%
Pipeline Fuel	31.8	37.4	17.6%	1.5%	1.7%
Other	33.0	36.3	10.0%	1.5%	1.7%
Total Transportation	**2,155.5**	**2,171.3**	**0.7%**		

Source: Historical emissions (1990 to 2006) from U.S. EPA (2008): *U.S. Inventory of Greenhouse Gas Emissions and Sinks: 1990 to 2006.* Projected emissions (2007 to 2030) from Cambridge Systematics analysis of Energy Information Administration, 2009: Annual Energy Outlook 2009 Updated April Release.

3.0 Greenhouse Gas Reduction Strategies and Impacts

3.1 STRATEGIES

Transportation greenhouse gas (GHG) emissions from fuel combustion and vehicle air conditioning systems account for 29 percent of total U.S. GHG emissions. In light of the aggressive national GHG reduction goals currently under discussion, which seek to reduce U.S. GHG emissions by as much as 80 percent from 2005 levels by 2050, the transportation sector could play a large role. The technical report for this study examines dozens of proposed strategies, and assesses their potential to reduce transportation GHG emissions. These assessments are based on published scientific literature, current policy studies, and best professional estimates. This section presents an overview and comparative summary of the technical report findings.

The strategies to reduce transportation GHG emissions discussed in the technical report are organized into four major groups. They include strategies to:

- **Introduce low-carbon fuels** Petroleum-based fuels account for 97 percent of U.S. transportation energy use.[61] The objective of this group of strategies is to develop and introduce alternative fuels that have lower carbon content and therefore generate fewer transportation GHG emissions. These alternative fuels include ethanol, biodiesel, natural gas, liquefied petroleum gas, low-carbon synthetic fuels (such as biomass-to-liquids), hydrogen, and electricity.

- **Increase vehicle fuel efficiency.** The objective of this group of strategies is to reduce GHG emissions by using less fuel per mile traveled. Fuel efficiency improvements include advanced engine and transmission designs, lighter-weight materials, improved aerodynamics, and reduced rolling resistance.

- **Improve transportation system efficiency.** These strategies seek to improve the operation of the transportation system through reduced vehicle travel time, improved traffic flow, decreased idling, and other efficiency of operations; improvements that can also result in lower energy use and GHG emissions. The strategies range from truck-idle reduction, to reducing congestion through Intelligent Transportation Systems (ITS) and other innovative forms of traffic management, to air traffic control systems that route aircraft more efficiently and reduce delays. Efficiency can also be

---

[61] U.S. Department of Energy, Annual Energy Outlook 2009.

improved by shifting travel to more efficient modes, where such shifts are practical in terms of price and convenience—such as passenger vehicle to bus, or truck to rail,

- **Reduce carbon-intensive travel activity.** The objective of this group of strategies is to influence travelers' activity patterns to shift travel to more efficient modes, increase vehicle occupancy, eliminate the need for some trips, or take other actions that reduce energy use and GHG emissions associated with personal travel.

The discussion begins with an overall comparison of the benefits and other impacts of these groups of strategies; noting the magnitude of reductions, as well as key issues related to cost-effectiveness, timing of benefits, and cobenefits. Key interactions among strategies and implications for infrastructure finance, are also discussed. This section concludes with a table detailing the benefits and impacts of each strategy.

Efforts that cut across these four strategy groups are addressed in Section 4. These are:

- **Pricing carbon.** Discusses pricing carbon through a cap and trade system, carbon tax, or increased motor fuels tax. The objective of this group of strategies is to reflect the broader costs of climate change by increasing the cost of emitting CO_2 and thereby influencing consumers and businesses to reduce CO_2 emissions. Policies that would price carbon emissions affect all four strategy groups: by increasing the cost of emitting greenhouse gases, they encourage the lowest cost combinations of the use of low carbon fuels, the purchase of energy efficient vehicles, the adoption of efficiency improvements in transportation systems, and the reduction of travel demand.

- **Transportation planning.** Discusses transportation planning and investment efforts that offer cross-cutting system efficiency and travel activity strategies. Transportation planning and investment decisions can improve the operating efficiency of the multi-modal transportation network, and integrate transportation and land use planning to reduce travel distances.

Pricing carbon is also summarized in this section for comparison with the GHG reduction benefits of other strategy groups. The GHG benefits of transportation planning and investment efforts could not be quantified and therefore are not discussed in this section.

3.2   ANALYSIS METHODS AND UNCERTAINTIES

Basis for Estimates

Estimates of GHG reductions as well as cost-effectiveness of the various strategies discussed here are based on material presented in Volume 2 of this report (or Volume 1, Section 4 for carbon pricing strategies). The analysis is based on a comprehensive review of the existing literature on the impacts of individual strategies. In some cases, additional original analysis was performed to develop estimates using a common set of assumptions and the most recently available data. The GHG reduction estimates for combinations of strategies were based on an assessment of what strategies could reasonably be implemented in combination with each other.

The benefits of strategies are presented for a "snapshot" 2030 analysis year, both in absolute terms (million metric tons of carbon dioxide equivalents, or mmt CO_2e), as well as in relative terms (percent reduction in total transportation GHG emissions compared to 2030 baseline projections). Total baseline transportation GHG emissions in 2030 are projected to be 2,171 mmt CO_2e, based on adjusted projections from the Department of Energy's Annual Energy Outlook (AEO) April 2009 release.[62] The AEO forecasts nearly level transportation GHG emissions over the next two decades, with emissions only 0.7 percent higher in 2030 than in 2007, as fuel efficiency increases offset increases in total travel.

For those strategies that would primarily have a long-term affect, 2050 impacts also are presented. Since the AEO does not include projections beyond 2030, the same baseline was used for 2050 as for 2030 (2,171 mmt) when calculating potential emission reductions.[63] This baseline would underestimate the potential emission reductions from these long-term strategies if in reality business as usual emissions would have been higher.

Sources of Uncertainty

Assessing the benefits of any particular strategy, or set of strategies, is a complicated and often controversial task that is best done at the time a strategy or set of strategies is being considered as a path forward. It is also critical to include in such an analysis the best, most current data and reliable assumptions. While the analysis presented in this report represents the most comprehensive assessment possible, given the existing scientific knowledge, there are many

---

[62] Minor adjustments to the AEO forecast were made to account for greenhouse gas emissions other than CO_2, as discussed in Appendix A. The AEO forecast does not account for additional increases in fuel economy targets beyond those established in the EISA, as are in the process of implementation by the Obama Administration in 2009.

[63] It is generally agreed that cumulative GHG emissions reductions over a future time period (for example, 2010 through 2050) are the most important measure of a strategy's success, rather than emissions in any particular year. Since cumulative emission benefits are not available for all the strategies in this report, however, common "snapshot" years are presented instead. The year 2030 is viewed as a reasonable "average" representation for the 2010-2050 period for strategies whose benefits increase over time (such as land use change or phase-in of new vehicle technologies).

inherent uncertainties. For example, some of the vehicle efficiency and low-carbon fuel technologies that could yield the greatest long-term benefits are still unproven. There is also no guarantee that these technologies can be advanced on a large scale to the point of feasibility and cost-competitiveness. Future fuel costs (as well as the relative difference in costs between fuels) will have a large impact on the cost-effectiveness of particular fuel-saving and alternative-fuel strategies, thereby affecting their ability to penetrate the market. For example, during periods of high petroleum prices, alternative fuels may be cheaper than gasoline, and vice versa. Therefore, both the magnitude and timing of these strategies should be considered uncertain.

In addition to technological uncertainty, many of the strategies from all categories may face significant political and/or institutional barriers. For example, some of these strategies may require greater up-front vehicle purchase costs (even if yielding net lifetime cost savings); significant public-sector investment; and may create negative impacts on some populations (even if yielding net social benefits). Carefully-crafted policies should seek to minimize any negative impacts of these strategies, and maximize the social benefits that are achieved beyond GHG reductions. Policy options for implementing GHG reduction strategies are discussed in Volume 1, Section 5.0 of this report.

For many strategies there is limited empirical evidence available to base findings. Often, only one or two studies have examined the GHG benefits of a particular strategy. A number of strategies had aspects that could not be evaluated due to this lack of empirical evidence. Even where multiple studies exist, they sometimes indicate a potentially wide range of benefits, reflecting uncertainty regarding the benefits or costs of these strategies. Professional judgment was applied in selecting the most comprehensive and reliable assessments to draw from for each strategy. In Volume 2, the level of confidence in the estimates, and any particularly important sources of uncertainty, are discussed for each strategy assessed.

Finally, there is uncertainty regarding consumer response to changes in travel conditions. Strategies that smooth traffic flow by reducing stop and go congestion may reduce emissions, but the improved travel conditions can also lead to increased travel, offsetting emissions benefits. While this concept, called *induced demand*, is widely acknowledged in the transportation profession, estimates of its magnitude are a source of uncertainty and debate. A range of plausible estimates from the literature could significantly impact induced demand and GHG calculations for many strategies. This study cites analysis that incorporates the impact of induced demand for strategies that would improve travel conditions–highway operations and management, public transportation, and commute travel reduction—with the exception of aviation, rail, and marine operations, where insufficient data was available.

For most strategies, a range of potential benefits is shown, reflecting some of the uncertainties inherent in the assessment. The ability to achieve even the lower end of the range, however, is by no means guaranteed. Therefore, the results presented here should be viewed as a general representation of the benefits that *could* potentially be achieved through each strategy, or group of strategies. The results also assume that advances in key technologies can be realized, and a favorable political and economic environment exists for the implementation of GHG reduction measures.

3.3 STRATEGY: INTRODUCE LOW-CARBON FUELS

This group of strategies introduces alternative fuels with lower carbon content per unit of energy than the current petroleum-based fuels powering the vast majority of today's transportation system. Examples of low-carbon fuels include ethanol, biodiesel, natural gas, hydrogen, and electricity. Fuel choices will depend on vehicle technology, since many fuels require some degree of powertrain modification, while others—such as electricity—require completely different powertrains. The current dominance of petroleum-based fuels reflects the advantages of liquid fuels for transportation, with their high energy densities allowing for extended vehicle range on limited storage. Gasoline and diesel in particular benefit from their firmly established production and distribution infrastructures, resulting in price advantages, and creating significant barriers to entry for most alternatives. Nonetheless, promising technologies that can provide low carbon fuel alternatives exist and are continuing to be developed.

The life-cycle greenhouse gas impacts of a fuel—not just emissions from the vehicle itself—must be considered when evaluating alternative fuel options. A life-cycle analysis takes into account the GHGs associated with all stages of the extraction (or feedstock production), processing, distribution, and dispensing of the fuel.

Natural gas can provide about a 15 percent GHG reduction relative to light-duty gasoline vehicles, roughly equivalent to diesel vehicle benefits. However, it requires more significant vehicle modification and distribution infrastructure, and its use is likely to be limited primarily to fleet vehicles utilizing central refueling and maintenance. Furthermore, its use as a transportation fuel would compete with use in other sectors, such as electricity generation and home energy, where it provides more cost-effective GHG reductions.[64]

Renewable fuels such as ethanol and biodiesel offer potential for GHG emission reduction. Renewable fuels are defined by the EPA as fuels produced from waste, plant or animal products, rather than fossil fuels. The GHG emissions benefits of renewable fuels depend on a variety of factors, including the feedstock, production method, carbon intensity of energy used in production, prior land use, and evaluation timeframe. Advanced biofuels from cellulosic sources will likely offer much steeper GHG reductions than first generation

---

[64] Vol. 2 Sec. 3.4.

biofuels, though more research and development is needed. Cellulosic ethanol is produced from the structural material that comprises much of the mass of plants.

Greater deployment of flex-fuel vehicles, which can run on either conventional or renewable fuel, and vehicles designed specifically to run on biofuels, would be needed to increase the market penetration of renewable fuels beyond 10 percent of light duty transportation fuel. Most vehicles on the road today can only operate on up to a 10 percent ethanol blend. Flex-fuel vehicles are slightly less efficient than vehicles designed to run on a single fuel. Ethanol has a lower energy content per gallon than gasoline, requiring slightly higher volumes of fuel. Adequate production capacity, land availability and distribution infrastructure are also key factors for renewable fuels.

Significant work is currently underway in the area of evaluating the effectiveness and cost of various renewable fuels in reducing GHG emissions. For example, the EPA published an analysis of life-cycle emissions from renewable fuels in conjunction with its revised renewable fuel standard. As such, this report to Congress does not include detailed analysis of biofuels. Readers are instead referred to EPA's renewable fuels website: http://www.epa.gov/otaq/fuels/renewablefuels/index.htm.

In the long-term (i.e., 25 years or more, with a projection year of 2050), hydrogen offers significant potential for GHG reductions, because hydrogen fuel cells are substantially more efficient than today's internal combustion engines. The GHG benefits of hydrogen depend strongly upon the method adopted for hydrogen production, but reductions per vehicle of about 50 percent in 2030 and 80 percent in 2050 could be realized with projected reductions in the GHG intensity of hydrogen production. However, hydrogen will only be a viable alternative if current technological barriers to fuel cells can be overcome. Furthermore, major investments in new production and distribution infrastructure will also be required to realize hydrogen's potential. Assuming these barriers can be surmounted, aggressive deployment could potentially lead to a 22 percent reduction in total transportation GHG emissions in 2050, if a 60 percent light duty vehicle market penetration could be achieved, which is the high end discussed in current literature. Production from renewable resources, or with carbon capture and storage (if this technology can be developed), will result in much greater benefits than production from non-renewable sources without carbon capture.[65]

Electricity shows similarly strong potential for GHG reductions, due to the inherent efficiency of electric motors. Electricity has the advantage of not requiring an entirely new production and distribution infrastructure. In addition, the cost of electricity for powering a vehicle is lower than that of

---

[65] Vol. 2 Sec. 3.8.

gasoline on a per mile basis. While vehicles using electricity generated from the current U.S. average generation mix can reduce GHG emissions by about 33 percent, compared to today's gasoline-powered vehicles, the GHG benefits of electricity will depend strongly upon the source of electricity generation.[66] Coal-fired electric plants may provide only modest benefits, or even increase net GHG emissions, unless successful carbon sequestration technologies are developed. Assuming increasingly lower GHG-intensity electricity generation, per-vehicle benefits could be as high as 78 to 87 percent in 2050, providing a total reduction in transportation emissions of 26 to 30 percent. Again, this assumes a 56 percent LDV market penetration by this time, which is the high end discussed in the literature. Considerable research and development on battery technology, notably to reduce costs and weight, is still required to bring electric vehicles to the point of being cost-effective and accepted by consumers.[67]

Cost-effectiveness is highly uncertain for most fuel options, and will depend upon advances in technology for the particular fuel and vehicle combination, as well as fuel prices. The costs for different fuels will fluctuate significantly depending upon supply and demand factors in other sectors, as well as the transportation sector, making it especially difficult to predict the market competitiveness of alternative fuels.

From a Federal policy perspective, several categories of policy options could be pursued in order to encourage adoption of low carbon fuels. Options include: fuel standards; market incentives, such as pricing and tax policies; and additional funding for research and development.

Other policy interventions may be helpful as well. For example, incentives for the production of flexible fuel vehicles can help overcome the dilemmas of fuel suppliers not introducing new low-carbon fuels until a sufficient number of vehicles can use them, and vehicle manufacturers not introducing alternative fuel-ready vehicles until a fuel supply is available.

For fuels entailing completely new production, distribution and vehicle platforms (e.g., hydrogen fuel cells), optimal modes or restricted markets can be identified to test comprehensive deployment on a small scale. Federal coordination and regulation may help ensure that both the fuel and vehicle sectors are focused on mutually supportive objectives.

---

[66] The 33 percent reduction may not be typical for in-use electric vehicles. It will depend upon the timing of when charging occurs, e.g., for peak vs. off-peak electricity demand periods, and furthermore vary by region of the country (depending upon the local generation mix).

[67] Vol. 2 Sec. 3.9.

3.4 STRATEGY: INCREASE VEHICLE FUEL EFFICIENCY[68]

Strategies to increase fuel efficiency are intended to reduce fuel consumption per unit of travel by on-road vehicles, locomotives, aircraft, and marine vessels. Although some retrofit options are available, energy efficiency improvements are inherently a medium- to long-range strategy, as they apply primarily to new vehicles and are dependent upon the rate of fleet turnover for the full affect on reducing transportation GHG emissions. Examples that are market-ready, and could be further incorporated in new vehicles in the near future, include: efficiency improvements to internal combustion engines; hybrid-electric powertrains and other efficiency improvements, such as weight reduction; and aerodynamic improvements that reduce drag. Many of these technologies are cost-effective, leading to net savings over the life of the vehicle, or even in a much shorter period, from reduced fuel consumption. In the longer term, entirely new propulsion systems relying on more efficient power conversion, and low- or zero-carbon fuels (such as hydrogen fuel cells), may be developed.

Some improvements to energy efficiency already are incorporated into baseline GHG projections, reflecting existing regulations and anticipated technological trends. For example, the efficiency of new cars and light trucks is projected in the AEO reference case to increase by 40 percent by 2030, as a result of expected impacts of CAFE standards through 2020, and then continue to increase beyond that time due to the effects of increasing fuel prices. Other sectors are not currently regulated for efficiency, but nonetheless are expected to show improvements (18 percent for aircraft, 14 percent for freight trucks, and 2 percent for rail and domestic shipping), as a result of technological advancements and market adoption.[69]

Improvements beyond this baseline show strong potential for further GHG reductions, with the largest potential in the light-duty vehicle sector. Estimates suggest that there is considerable potential for improvement beyond what will be achieved by the CAFE standards established under the 2007 EISA.

Potential GHG reduction benefits per vehicle (compared to the AEO baseline projection for conventional gasoline vehicles) in 2030 and beyond range from 8 to 30 percent for advanced conventional gasoline vehicles; 0 to 16 percent for diesel vehicles; 26 to 54 percent for hybrid-electric vehicles (HEVs); 46 to 70 percent for plug-in hybrid electric vehicles (PHEVs); 40 to 84 percent for hydrogen fuel cell vehicles; and 68 to 87 percent for battery-electric vehicles as shown in

---

[68] Vol. 2 Chapter 3

[69] The truck, air, rail, and marine efficiency measures are based on energy use per ton-mile (freight movement) and seat-mile (aircraft), and therefore may be affected by utilization (load) factors as well as inherent vehicle efficiency. Source: AEO 2009 Reference case, April 2009 release.

Figure 3.1.[70] Key fuel technologies that rely on new vehicle technology (hydrogen fuel cell and battery-electric) are also presented in Figure 3.1 for comparison.

Figure 3.1   Projected Future GHG Benefits of Light-Duty Vehicle/Fuel Technologies Compared to Baseline Conventional Gasoline Vehicle

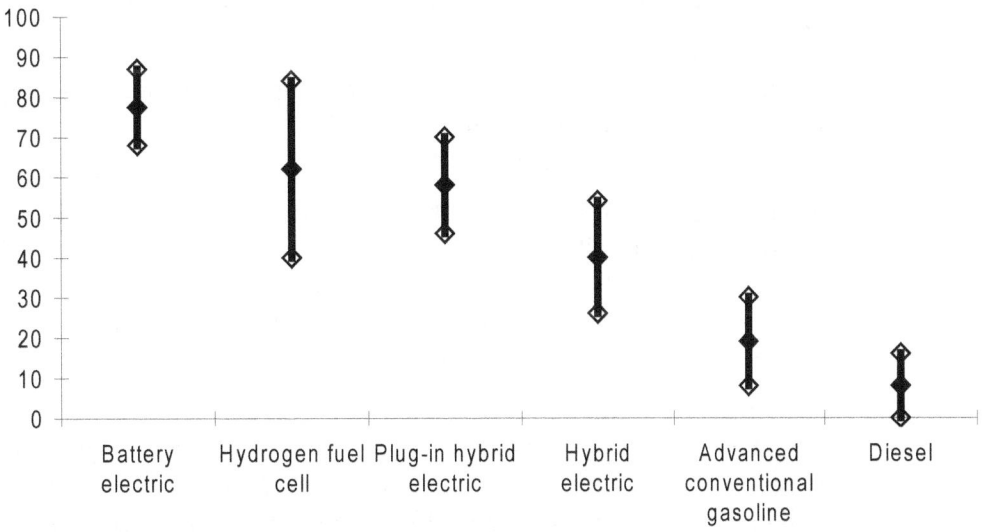

Percent Reduction in Life-Cycle GHG versus Conventional Gasoline

Source: Eastern Research Group, Inc. analysis as presented in Vol. 2, Sec. 2.8 and 2.9 (hydrogen fuel cell and battery electric) and Vol. 2, Sec. 3.2 (other vehicle types). The ranges shown represent GHG reductions for 2030 and beyond, with the low end of the battery electric and hydrogen fuel cell ranges reflecting 2030 impacts and the high end reflecting additional advances through 2050.

The estimates for plug-in hybrid and battery electric vehicles depend on reductions in the GHG emissions intensity of U.S. electricity production. The estimates were calculated using GHG emission intensity modeled by the Electric Power Research Institute (EPRI).[71] The input is 379 to 606 g/kWhr in 2030, and 240 to 421 g/kWhr in 2050. This compares to a 618 g/kWh national average today and would require increased use of low carbon electricity production technologies such as wind, solar, nuclear, and hydro-electric power. However, even under a very high GHG intensity scenario relying on coal generation using

---

[70] Vol. 2 Sec. 3.2.

[71] Electric Power Research Institute (EPRI). *Environmental Assessment of Plug-In Hybrid Electric Vehicles. Volume 1: Nationwide Greenhouse Gas Emissions.* Report no. 1015325. 2007.

older technology (1,014 g/kWhr), at a low battery efficiency of 0.4 kWhr/mile, PHEVs operating in a charge depleting mode would still result in 12 percent lower GHG emissions than corresponding conventional gasoline vehicle operation, on a per mile basis. However, under these extreme circumstances, PHEV operation will not provide benefits relative to an HEV baseline.[72]

Retrofits can be used to speed improvements. Retrofits of heavy-duty trucks to use aerodynamic fairings, trailer-side skirts, low-rolling resistance tires, aluminum wheels, and planar boat tails reduce per truck GHG emissions by 10 to 15 percent. For new trucks, combined powertrain and resistance reduction technologies are estimated to reduce per vehicle emissions by 10 to 30 percent in 2030.

Aircraft advances, including open-rotor engines or blended-wing designs, could potentially reduce GHG emissions by 10 to 40 percent per aircraft from baseline conditions, as these technologies are phased in over the next 20 to 30 years.[73]

Significant improvements in the efficiency of rail and marine vehicles are also possible—potentially 20 percent or more with an effective suite of advanced technologies. However, these sectors' contributions to total transportation emissions are relatively small, and the total GHG reductions that may be achieved are somewhat less than for other sectors.[74]

The emission reductions associated with vehicle improvements will start slowly, but increase over time as new technology is phased in. Fleet turnover varies by type of vehicle, occurring more quickly in the light-duty sector (where most vehicles are replaced within 15 years), than for trucks, railcars, and marine vessels (for which vehicle lifetimes typically range from 20 to 40 years); and about 30 years for aircraft. Despite the long lifetimes of these vehicles, newer vehicles tend to be used more than older vehicles (especially in the trucking sector), and therefore GHG reduction benefits will occur somewhat more quickly. Some near-term reductions may also be obtained through retrofits of trucks, railcars and marine vessels.

Technologies to improve vehicle fuel efficiency generally have the effect of increasing the initial purchase price of the vehicle, or requiring up-front capital investment in retrofits. However, these technologies also yield cost-savings over time due to reductions in fuel consumption. Many—including most advanced gasoline vehicle technologies, truck efficiency improvements, and rail and marine technologies--yield net cost-savings over the life of the vehicle and can cover the initial investment within a much shorter timeframe. Others, such as diesel, hybrid-electric, and plug-in hybrid electric vehicles, have somewhat more

---

[72] Vol. 2 Sec 3.2.

[73] Vol. 2 Sec. 3.7.

[74] Vol. 2, Sec. 3.5 and 3.6.

uncertain lifetime cost-effects. In general, fuel efficiency technologies will become more cost-effective as the price of fuel increases. Other important factors influencing cost-effectiveness include the relative costs of different fuels and future battery costs for plug-in hybrid-electrics.[75]

A range of Federal policy initiatives can influence the rate of technology advancement and the adoption of high-efficiency technologies. Vehicle- and fuel-related policies should be considered simultaneously to maximize the effectiveness of these policies and ensure that unnecessary overlap or redundancy among policies does not occur.

Broadly, these approaches can be categorized as:

- GHG efficiency and fuel economy standards for new vehicles, such as the NHTSA and EPA harmonized National Program;

- Partnerships with industry to develop standards and demonstrate new technologies;

- Subsidies or tax credits for efficient vehicles or retrofits; either for new vehicles or for existing vehicles, as annual payments through the registration process;

- Updating tax rate on inefficient vehicles (i.e. Energy Tax Act of 1978 or "gas guzzler tax"), levied either on new vehicles, or as annual fees on existing vehicles; and subsidies for scrapping the most inefficient vehicles (http://www.epa.gov/fueleconomy/guzzler/);

- Fuel taxes, VMT fees, or "cap and trade," through their effect on fuel prices and operating costs; and

- Research and development subsidies.

The effectiveness and desirability of such programs will depend, in part, on whether or not there are market failures that cause firms to fail to develop efficient vehicles, or that would deter users of such vehicles from purchasing more efficient vehicles. In the case of light duty vehicles, for instance, it has been argued that consumers, for various reasons, do not fully consider future fuel costs in their purchases of new vehicles.[76] In any case, private purchasers of new vehicles would not normally consider any external public social costs of climate change or petroleum imports. While market failure arguments are less compelling for manufacturers and purchasers of expensive commercial transportation equipment—such as airliners, locomotives, and ships—higher

---

[75] Vol. 2, Sec. 3.

[76] Greene, D. L., J. German and M. A. Delucchi (2009). "Fuel Economy; The Case for Market Failure." In *Reducing Climate Impacts in the Transportation Sector*, D. Sperling and J. S. Cannon, eds, Springer.

capital costs (as well as the investment required to develop new technology) can still provide a barrier to adoption of new technology; given the risk created by uncertainty over future fuel prices.

Many of the most promising technologies for increasing efficiency still face significant technological hurdles (such as fuel cells with its concomitant hydrogen infrastructure challenges), or substantial cost and performance disadvantages (such as electric batteries). Federal funding for research and development for vehicles of all modes could help overcome these hurdles. More stringent vehicle efficiency regulations would encourage private-sector investment in research and development, as would substantial and sustained increases in fuel or carbon prices. Vehicle efficiency regulations would have the greatest impact in the light-duty vehicle sector, where fuel costs are a relatively small factor in consumers' vehicle purchase decisions. Other technology-neutral incentives, such as "feebates" that increase or decrease a vehicle's purchase cost—depending upon its relative energy efficiency—could serve as an alternative or supplement to efficiency regulations. Finally, the Federal government could potentially adopt standards for technologies that are proven to be cost-effective (e.g., for heavy-duty vehicles), or work with international organizations, such as the United Nations' International Civil Aviation Organization and International Maritime Organization, to adopt standards for marine vessels and aircraft.

3.5 STRATEGY: IMPROVE TRANSPORTATION SYSTEM EFFICIENCY

Strategies to improve transportation system efficiency seek to optimize the use of the transportation network by improving transportation operations and reducing energy use and GHG emissions associated with a given unit of passenger or freight travel (e.g., person-miles, vehicle-miles, or ton-miles). The collective impact of these strategies is relatively modest compared to vehicle and fuel technology strategies—approximately a 3 to 6 percent reduction relative to baseline 2030 transportation emissions.[77] Unlike vehicle and fuel technology strategies, however, many of these strategies also have significant co-benefits in the form of time-savings to travelers and reduced costs to shippers. Furthermore, they may represent important GHG reduction strategies on a local basis (e.g., in highly congested areas).

System efficiency strategies rely largely on the planning, design, operations, and management of transportation systems--factors within the control of national, state, and local transportation agencies. Efficiency strategies, such as intelligent traffic management, can lower GHG emissions by reducing fuel consumption

---

[77] Vol. 2 Sec. 4.1.

associated with congestion (estimated at nearly 3 billion gallons per year[78]). Operational efficiencies such as idle reduction, delay reduction, and more efficient routing and scheduling can also achieve benefits in the truck, rail, aviation, and marine sectors.

There are several sources of uncertainty in calculating the GHG benefits of system efficiency strategies, and especially those that reduce congestion. Most significantly, the benefits of both highway and air improvements may be offset by induced travel demand resulting from lower travel times and costs (see sidebar on p. 3-21). Second, the total GHG reduction benefits will decrease over time if vehicle fuel efficiency increases beyond projected baseline levels, or the carbon content of fuels decreases. Thirdly, construction projects result in additional greenhouse gas emissions from the operation of related equipment and traffic delays during the construction process, but these effects have not been rigorously quantified and are not included in existing GHG estimates for these strategies. Because of these uncertainties, numerical estimates are not included for highway operations and investment strategies.

Characterization of GHG Reductions Used in this Report:
In this report, when referring to individual strategy effects, "**modest**" refers to reductions in CO_2e emissions of less than 0.5 percent of total transportation emissions, or 12 mmt in 2030; "**moderate**" to reductions in the range of 0.5 to 2.5 percent of total transportation emissions, or 12 to 60 mmt in 2030; and "**significant**" to reductions of greater than 2.5 percent of transportation emissions or 60 mmt in 2030.

Highway traffic management strategies and real time traveler information, including signal timing, freeway ramp metering, faster clearance of incidents, and variable message signs, have modest potential for reducing GHG emissions; even if induced travel demand from these projects is considered, and presuming that these projects do not result in substantial GHG emissions from project construction. Outside analysis suggests that widespread deployment of these intelligent traffic management strategies could produce modest GHG benefits by reducing inefficient vehicle operations.[79]

---

[78] *See* "What Does Congestion Cost Us?" in *2009 Urban Mobility Report*, published by the Texas Transportation Institute.

[79] By way of example, the *Moving Cooler* analysis used FHWA's Highway Economic Requirements System (HERS) model and its embedded assumptions regarding induced demand to estimate the impacts of traffic management strategies. This analysis suggested that these strategies could reduce total transportation GHG emissions by as

Footnote continued

While their GHG benefits may be modest, highway traffic management and traveler information strategies have significant co-benefits, especially in the form of time-savings to travelers, as well as the economic benefit of cost-savings for shippers. Traffic management strategies such as signal coordination and incident management are proven, have relatively modest costs, and could be more broadly deployed within the next 5 to 10 years—yielding early GHG reductions that may be significant at a local scale.

Highway bottleneck relief strategies involve increasing capacity at "bottlenecks" (specific points on the transportation network where demand exceeds capacity), through such measures as added lanes, interchange improvements, and intersection reconfigurations. Outside analysis from the *Moving Cooler* study shows modest GHG reductions from bottleneck relief strategies in 2030, but modest increases in GHG emissions by 2050, because of induced demand. [80]

A Federal policy to **reduce speed limits** (for example, from 70 to 60 mph or from 65 to 55 mph) on national highways would generate substantial immediate benefits, reducing total transportation GHG emissions by 1.1 to 1.8 percent; in addition to having significant safety benefits. However, achieving these benefits would require strong enforcement, and by reducing travel speeds this strategy would increase travel times, and could increase costs to shippers. Stronger Federal funding incentives and disincentives, coupled with Federal oversight, would be required to achieve more effective enforcement if this strategy is pursued. This strategy is quite cost-effective, with enforcement costs of about $10/tonne GHG reduced. [81]

---

much as 0.6 percent in 2030 (see Vol. 2, Sec. 4.2.1). The *Moving Cooler* estimates also showed net GHG reductions in 2050.

[80] The *Moving Cooler* study analyzed the impact of bottleneck relief construction projects at the top 200 bottlenecks in the United States. It found that bottleneck relief strategies would achieve a net reduction in GHGs of 4 mmt CO2e under maximum deployment in 2030 and a net increase in GHGs of 10 mmt CO2e under maximum deployment in 2050. That corresponds to a 0.3% decrease in US on road GHGs in 2030 and a 0.7% increase in US on road GHGs in 2050. These estimates do not include construction emissions (see Vol. 2, Sec. 4.2.3). The bottleneck relief estimates also assume that the projects would be fully financed by increased fuel taxes, which somewhat mitigates the induced travel demand resulting from congestion relief.

[81] Vol. 2 Sec. 4.2.4.

Induced Travel Demand

Induced travel demand can be defined as any increase in travel resulting from improved travel conditions. The induced VMT generally results from longer trips, as well as additional trips and shifts of travelers from other modes. Over the longer term, improved travel conditions can also impact land use, further impacting trip lengths and modal shifts. It is an important consideration for system efficiency and travel activity strategies, affecting the impacts on travel and corresponding GHG benefits of most of these strategies.

In particular, bottleneck relief, traffic management, and traveler information strategies lead to additional travel by reducing congestion and travel times; this additional travel reduces and, in the long run, potentially eliminates the effectiveness of these measures in reducing GHG emissions. To a lesser extent, travel behavior strategies that reduce on-road trips also result in induced demand, since the initial reduction of highway travel times will draw some additional traffic back onto these facilities. Induced demand is related to the basic economic concept of elasticity, meaning that a decrease in cost (such as travel time) results in an increase in consumption. Sources referenced in this report applied short- and long-term elasticities to estimate induced demand effects, and used adjusted travel volumes to calculate fuel consumption and GHG emissions. Strategies that reduce VMT by making highway travel more expensive – such as mileage-based fees, congestion-based tolls, or increased gas taxes – are assumed to result in no induced demand, since the increase in monetary costs suppresses the demand for additional travel. Use of "congestion pricing" in connection with bottleneck relief strategies may limit offsets from induced demand.

While the concept of induced demand is widely acknowledged in the transportation profession, estimates of its magnitude are a source of uncertainty and debate. A range of plausible estimates from the literature would significantly impact induced demand and GHG calculations for many strategies. U.S. DOT is designing research to provide a better understanding of the role of induced demand in offsetting GHG improvements from congestion reduction strategies.

This study used the same induced demand assumptions as those recently used in the Federal Highway Administration's (FHWA) Highway Economic Requirements System (HERS) model. The version of HERS used for the 2008 U.S. DOT Conditions and Performance Report uses an elasticity of VMT with respect to total travel cost of -0.4 for the short run and -0.8 for the long run. To compare an elasticity for fuel prices to an elasticity for total travel costs, one would need to multiply the fuel price elasticity by a factor of three to ten, since fuel cost represents only about a tenth to a third of total operating costs. Small and Van Dender (2007) estimate an elasticity of VMT with respect to fuel prices of between -0.02 and -0.03 for the short-run and of an elasticity between -0.11 and -0.15 for the long-run. For short- to medium-run responses of VMT to changes in fuel prices, Ewing et al. (2008) estimated an elasticity of -0.17. The question of how strongly VMT responds to changes in travel costs is far from settled, with ongoing research continuing to produce new estimates. Additional details on the calculations performed by the sources cited in this study can be found in Section 4.1.4 and 4.2 of Volume 2 and Appendix A.

Traveler information provides up-to-date information to travelers on traffic conditions, incidents, and expected delays; the availability of public transportation and other travel alternatives; weather conditions; road construction; and special events. While valuable in improving the timing and routing of travel choices, it provides only modest GHG benefits; at least in the current form of real-time road traffic information. Traveler information can be deployed with modest resources. Additional benefits may be realized in the future through new strategies, such as real-time rideshare matching and transit information. However, reliable information does not yet exist on the travel and GHG impacts of these emerging strategies.

Truck idle reduction provides only modest GHG benefits—up to 0.2 percent of total transportation GHG--but could be implemented relatively quickly and provides net cost savings to vehicle operators (with a short payback period of two to three years). It also reduces local air pollutants. Truck idle reduction could be implemented through the adoption of a uniform national anti-idling law, in combination with financial incentives for the purchase of idle reduction technology.[82]

For the **rail and marine sectors**, efficiencies can be achieved through rail chokepoint relief to reduce congestion, as well as revised operational practices, such as locomotive idle reduction in rail yards and shore-side power use for ships. The EPA's 2008 rulemaking includes requirements to reduce emissions from idling locomotives by requiring technology that reduces the amount of time a locomotive spends idling and applying tighter emission standards to new locomotives. These efficiency improvements provide modest benefits in GHG reductions from operations and may also encourage the shifting of freight from trucks to the more efficient rail and marine modes. The potential for freight mode shifting is limited by many factors, including haul distance (most efficiency benefits are lost for shipments less than 500 to 1,000 miles), handling costs at terminals, and the demand for speed and reliability in the shipment of high value or time sensitive freight. The collective potential reduction of transportation GHG emissions from rail and marine operations appears to be less than 0.5 percent of all transportation GHG emissions, with most of the potential in the rail sector. While some rail and marine operating strategies can be implemented at a modest cost, nationwide elimination of key rail chokepoints requires substantial private and public investment.[83] Improvements to intermodal operations, such as reducing chokepoints where freight is transferred between marine, rail, and highway modes, can also reduce emissions.[84]

---

[82] Vol. 2 Sec. 4.3.1.

[83] Vol. 2 Sec. 4.4.

[84] Vol. 2 Sec. 4.4.

Improvements to **aviation efficiency,** such as more direct routing and more efficient takeoff and landing profiles, show the potential to increase air traffic operational efficiency by 2.5 to 6 percent by 2035. Many of these improvements already are being implemented through the FAA's NextGen program. Other operational improvements at airports (e.g., single-engine taxi, electric gate power) show very modest potential GHG benefits, although they may have significant co-benefits in the form of reductions in local air pollution and airline cost savings. Aviation efficiency improvements that reduce the cost of air travel could potentially result in offsetting increases in GHG emissions as more people travel, but this effect has not been reliably quantified and is not included in the estimates presented here.[85]

Transportation infrastructure construction is a significant contributor to GHG emissions. These emissions are discussed under life-cycle emissions in Section 2.3, since only tailpipe emissions are included as transportation emissions in the U.S. GHG inventory for accounting reasons. Perhaps the most significant and currently available strategy to reduce GHG emissions from construction is the use of fly ash in cement, which uses a recycled material to reduce the amount of cement needed by up to 50 percent (cement production produces large amounts of GHGs). Already in use in a few places, this strategy could be implemented much more widely as State DOTs become more comfortable with the technology. Greater widespread use of warm- and cold-mix asphalt also has the potential to reduce GHGs generated to produce and laydown these asphalt materials, but further research and demonstration under a variety of conditions in the U.S. is needed. Together these strategies have the potential to reduce GHG emissions by roughly 0.8 percent, relative to the transportation sector baseline in 2030.[86]

3.6 STRATEGY: REDUCE CARBON-INTENSIVE TRAVEL ACTIVITY

Strategies to reduce carbon-intensive travel activity seek to influence travelers' patterns in order to shift travel to more efficient modes, increase vehicle occupancy, reduce the need for travel, or otherwise take actions that reduce energy use and GHG emissions associated with personal travel. The collective impact of these strategies on transportation GHG emissions could range from 5 to as high as 17 percent in 2030; or 6 to 21 percent in 2050.[87] The greatest near-

---

[85] Vol. 2 Sec. 4.5.

[86] Vol. 2 Sec. 4.6.

[87] Vol. 2 Sec. 5.1. Some of the benefit estimates for a number of travel activity strategies, including transit, nonmotorized improvements, land use, and commuter strategies, incorporate "induced demand" effects. As some travelers shift to other modes or reduce

Footnote continued

term benefits could come from pricing strategies such as "pay-as-you-drive" insurance, as well as "eco-driving" training and in-vehicle equipment to encourage more efficient driving techniques. In the long-term, substantial benefits may be realized from changes to land use and transportation infrastructure (such as transit and nonmotorized investment) to reduce trip distances and support greater utilization of more efficient travel modes. By providing travel alternatives and enabling shorter trips, these strategies can increase access to jobs and other economic opportunities.

Pricing strategies have significant potential to reduce GHG emissions within a short timeframe, as consumers respond directly to price signals and adjust their travel patterns accordingly. Comprehensive pricing strategies that affect all travel — such as higher motor fuel taxes, VMT fees, or pay-as-you-drive insurance — could reduce GHG emissions by 0.7 to 3.1 percent within 5 to 10 years. This is based upon an estimated pricing implementation of an additional 2 to 5 cents per mile, which is roughly equivalent to a $0.40 to $1.00/gallon gas tax.[88] Pay-as-you-drive would actually reduce costs for a majority of travelers, although it would also increase it for some. Strategies focused on specific markets, such as inter-city tolls or cordon pricing, would have more limited benefits consistent with the size of the market affected. Widespread congestion pricing, in which higher prices are charged for traveling in periods of high demand, would not only reduce VMT but also result in more efficient traffic operations. The Federal government could encourage pricing strategies through a number of mechanisms, such as: requiring states to allow pay-as-you-drive insurance; implementing a nationwide VMT fee; providing funding incentives or disincentives for states or metropolitan agencies to implement pricing mechanisms; allowing expanded tolling on Federal-Aid highways; or increasing the Federal motor fuel tax. In order for pricing to yield net benefits to the traveling public, and not produce unacceptable equity impacts, revenues from road pricing would need to be reinvested in services that benefit effected travelers (such as critical transportation asset State of repair, transit, land use planning, and other strategies that improve accessibility) or returned to taxpayers.

Nearly two-thirds of VMT occurs in urban areas. Expansion of **urban transit** has the potential to generate modest to moderate reductions in GHG emissions. Under the scenario of investing in transit sufficiently enough to nearly double the average annual ridership growth rate (from the current 2.4 percent to 4.6 percent), expanded urban transit could reduce GHG emissions from 0.2 to

---

their travel, roadway congestion will be reduced, thereby potentially allowing or encouraging other people to drive more. This effect has been estimated to reduce GHG benefits of these strategies by a modest amount (about 14 percent) as discussed in Appendix A.

[88] Vol. 2 Sec. 5.2.

0.9 percent of transportation GHG by 2030, or 0.4 to 1.5 percent in 2050.[89] Benefits would increase over the long-term as transit service, connectivity and reliability increase. While transit expansion is costly – over $1,000 per ton when considering transit capital investment and operations costs – it can result in significant co-benefits to travelers in the form of improved mobility, especially for low-income travelers. Transit expansion would also result in cost savings for personal vehicle ownership and operation, with net savings of up to $900 per ton when these costs are included.[90] **Inter-city transit**, including high-speed rail and bus, also has the potential for GHG reduction – up to 0.2 percent of transportation emissions in 2030.[91]

Non-motorized improvements, including construction of pedestrian and bicycle transportation networks through dedicated rights-of-way, as well as enhancements to existing rights-of-way that safely provide for bicycle and pedestrian traffic, have modest potential for GHG reductions. These measures would reduce GHGs by 0.2 to 0.6 percent by 2030, at moderate investment costs (less than $200 per ton), or a net savings when reduced vehicle operating costs are considered. While their GHG benefits may be modest, these strategies also provide significant cobenefits in the form of improved livability as well as mobility for travelers who do not drive.[92] These improvements, especially those for pedestrian mobility, are closely linked to land use changes discussed in the next section that describe how our residential housing, transportation, and other infrastructure choices are linked.

Land use changes -- such as density, diversity of land uses, neighborhood design, street connectivity, destination accessibility, distance to activity centers, and proximity to transit -- reduce trip lengths and support travel by transit, walking, and bicycling. This report to Congress analyzed the literature to develop a range of potential GHG reductions from land use strategies. Three studies were particularly instructive: *Growing Cooler*, authored by academic and industry researchers and published in 2008 by the Urban Land Institute; *Moving Cooler*, authored by Cambridge Systematics and published by the Urban Land Institute in 2009; and *Transportation Research Board Special Report 298: Driving and the Built Environment*, published by the National Academy of Sciences in 2009. All three studies, conducted independently and using different assumptions and analysis methods, found GHG reductions from land use strategies of the same order of magnitude. Taking the middle section of the study ranges and adjusting

---

[89] This scenario would involve a capital investment of approximately $71 billion over the 2006 – 2026 period, compared to $42 billion to accommodate current levels of ridership growth; see Vol. 2 Sec. 5.3.1.

[90] Vol. 2 Sec. 5.3.1.

[91] Vol. 2 Sec. 5.3.2.

[92] Vol. 2 Sec. 5.3.3.

them to the same baseline as that used in this report to Congress, yields a reduction of U.S. transportation GHG emissions of 1 to 4 percent in 2030 and 3 to 8 percent in 2050.[93] The *Moving Cooler* study assumptions, which fall in the middle of the range, rely on 43 to 90 percent of new urban development occurring in areas of roughly greater than five residential units per acre, which accommodates single family and multifamily homes.[94] It does not assume changes in rural development. GHG reductions from land use change increase over the long term, as land use patterns evolve over long periods of time due to the resilience of the existing housing stock and transportation infrastructure. Transit, nonmotorized improvements, and pricing would be most effective over the long term if they are implemented in combination with more compact and better integrated land use patterns that reduce overall trip lengths and make alternative modes viable as a means of travel for many trips. Land use changes can often be implemented with very little public investment cost, with the primary direct costs including knowledge sharing, outreach, and planning activities. Additional infrastructure investments, or other costs such as brownfields cleanup, may be needed in some locations, but on the whole more compact land use patterns have been demonstrated to provide long-term cost savings through reduced roadway and other infrastructure requirements. While land use planning is conducted at the local level, the Federal government could encourage changes to land use patterns by funding State and regional planning activities to coordinate local policies; building State and local capacity to understand, model, and assess sustainable development principles within project planning; and by providing incentives and/or disincentives through transportation funding mechanisms.

Commuter/worksite trip reduction programs have modest potential for GHG reductions—0.2 to 0.6 percent of all transportation sector emissions in 2030. The most effective actions from a policy perspective are trip reduction requirements combined with supporting activities such as regional rideshare and vanpool programs and financial incentives for the use of alternative modes. Federal funding for aggressive public outreach programs to encourage employers to offer travel alternatives could be effective even in the absence of mandates. Telework and other alternative work schedules can further reduce GHG from work travel by up to 0.5 percent, although telework is likely to spread largely through private initiative and the role of the public sector in encouraging adoption of alternative work schedules appears limited.[95]

Most **public information campaigns** exhibit modest GHG reduction potential— in the range of 0.1 to 0.2 percent of transportation GHG emissions, although most

---

[93] Vol. 2 Sec. 5.4.

[94] For visuals of different density levels, please see Vol. 2 Sec. 5.4.

[95] Vol. 2 Sec. 5.5.

can be implemented quickly. Campaigns based on mass marketing have demonstrated little ability to influence travel behavior. Individualized marketing, in which people are provided with customized information on travel alternatives, shows somewhat greater promise in areas where good alternative services are available. Educational efforts to encourage eco-driving and proper vehicle maintenance have shown some short-term benefit, but the impacts tend to diminish over time. More comprehensive and sustained efforts to promote eco-driving, including requiring instruction as part of driver education and providing in-vehicle feedback technology, could reduce transportation GHG emissions by up to 1 to 4 percent, although findings on eco-driving benefits are based on limited European experience that may not be replicable in the United States.[96]

3.7 STRATEGY: PRICE CARBON

Pricing carbon through a cap and trade system, carbon tax or increased motor fuels tax would affect vehicle fuel efficiency, encourage use of low-carbon fuels, and encourage more energy-efficient travel patterns.

Either a cap and trade system or a carbon tax approach would create a consistent set of prices across all sectors to encourage actions to reduce GHG emissions. Within the transportation sector, these actions would increase the cost of carbon-fueled transportation and would therefore create incentives for developing and purchasing more efficient vehicles and alternative fuels, as well as reducing travel and/or shifting to more efficient modes. An increase in the Federal motor fuels tax produces the same effects for transportation modes that use gasoline and diesel fuels. The longer-term impact on fuel consumption and GHG emissions would be greater than the immediate impact, as transportation system users, fuel providers, and vehicle manufacturers have time to respond with changes to vehicles, fuels, and basic activity patterns.

Analysis by the Energy Information Administration (EIA) of the cap and trade system in H.R. 2454, the American Clean Energy and Security Act (ACES)[97] found reductions in transportation GHG emissions from ACES of about 4 percent in 2030 relative to baseline emissions, or 85 million metric tons CO_2e.[98] This reduction results from a gasoline price increase of about 37 cents per gallon

---

[96] Vol. 2 Sec. 5.6.

[97] Passed the House of Representatives in June 2009 but a companion Senate bill has not passed as of this writing.

[98] U.S. Department of Energy, Energy Information Administration (2009). *Energy Market and Economic Impacts of H.R. 2454, the American Clean Energy and Security Act of 2009.* http://www.eia.doe.gov/oiaf/servicerpt/hr2454/index.html. Figures cited here are for the basic case.

in 2030, corresponding to a $65 per tonne allowance price.[99] A carbon tax instituted at a comparable level to the permit price of a cap and trade system would have similar impacts. Increasing the Federal motor fuels tax would also have a similar impact, but would only raise prices on gasoline and diesel, rather than applying to all fuels based on carbon content.[100] This section does not examine the impact of carbon pricing on aviation or maritime industries.

Costs to the broader economy of cap and trade proposals are estimated on the order of less than one percent of U.S. gross domestic product (GDP) in 2030 and two percent of GDP in 2050. Both a carbon tax and a cap and trade system could be made more socially equitable by, in the case of a carbon tax, giving rebates to low income households, and in the cap and trade system, compensating low income households using a portion of the revenue from the auction of allowances. [101]

3.8 KEY INTERACTIONS

Many of these strategies interact to produce different outcomes in total GHG reductions. The benefits of each strategy (or group of strategies) are not additive, and in fact may be reduced depending on other strategies that are implemented. On the other hand, some strategies are complementary and their effectiveness is likely to be enhanced if implemented in combination with each other. As examples:

- The effect of market mechanisms and vehicle efficiency standards would be somewhat overlapping. An increase in the cost of carbon should provide incentives for the development and purchase of more efficient vehicles as well as for reducing travel. The GHG reductions from vehicle efficiency improvements would therefore be the maximum of those caused by the fuel price increase or those set by regulatory standards—not the sum of the impacts if either were applied individually. With higher CAFE standards already in place, the most cost-effective fuel efficiency technologies would already be adopted, meaning that the additional benefits of modestly higher fuel prices under a cap and trade system are small. A recent U.S. DOE analysis of proposed cap and trade legislation found that additional fuel efficiency improvements would be very modest—about 1.2 percent for light-

---

[99] The modeling performed by EIA finds that gasoline prices change under a policy scenario not only because of the direct impact of the allowance requirement, but also because of general equilibrium effects, such as a lower demand for fuels leading to slightly lower world crude oil prices.

[100] Vol. 1 Sec. 4.1.

[101] Vol. 1 Sec. 4.1.

duty vehicles in 2030 .[102] On the other hand, if higher fuel prices are sustained over time (either because of market forces, or because of dramatic fuel or carbon tax increases) the additional benefit of technology standards would be lessened. Furthermore, economy-wide pricing would have impacts on other modes (albeit quite modest, at fuel price levels predicted under cap and trade proposals—see Section 4.1) that are not affected by energy efficiency regulations. An increase in the price of fuel or carbon would also have the effect of shifting vehicle purchases between segments of the light-duty vehicle market, i.e., from light trucks to cars, as was seen when gas prices spiked in 2008 and sales of SUVs dropped while demand for fuel efficient cars rose. CAFE standards may or may not have this effect. Separate standards for light trucks and cars could decrease purchase shifts. However, some manufacturers significantly discount fuel efficient models in order shift more of their sales to these models and meet CAFE standards.

- Some vehicle and fuel strategies are interrelated. Only a few fuels—notably, biodiesel and ethanol at blends of 10 to 15 percent or less—can be used directly in today's vehicles without modification. Most low-carbon fuels such as higher ethanol blends or natural gas, require at least minor modifications to vehicle design. Some, notably electricity and hydrogen, benefit from or require the development of entirely new vehicle propulsion technologies.[103] Furthermore, the total benefits cited for fuel efficiency and low-carbon fuel strategies are not additive. To determine the total benefits from these strategies, it would be necessary to construct scenarios of future market penetration for different vehicle and fuel technology combinations. From a policy perspective, the Federal government can play an important role in ensuring that research and development activities, regulations, and infrastructure deployment are coordinated to promote a complementary set of vehicle and fuel technologies.

- As vehicle efficiency increases and/or fuel carbon content decreases, the absolute GHG reduction benefits of system efficiency and travel activity strategies (such as signal coordination, pricing, land use, and transit) will decrease proportionately. Furthermore, some vehicle technologies will reduce the benefits from system efficiency strategies aimed at addressing congestion or idling, as GHG emissions associated with congestion or idling are minimized through the use of fuel cell, electric, and hybrid electric-

---

[102] U.S. Department of Energy, Energy Information Administration (2008). "Energy Market and Economic Impacts of S. 2191, the Lieberman-Warner Climate Security Act of 2007."

[103] In addition, diesel vehicles – treated as a vehicle technology strategy in this report – could easily be considered a fuel strategy; and plug-in hybrid vehicles, also treated in vehicle technology, also make use of electricity as a fuel source and share many of the same characteristics as battery-electric vehicles.

drivetrains. For example, hybrid-electric vehicles typically achieve fuel economy on urban driving cycles that is close to or exceeds fuel economy on highway cycles. System efficiency is still a valuable goal for other reasons, however, including improving mobility, reducing congestion and delay, and reducing shippers' costs.

- Transit, nonmotorized improvements, land use, and pricing strategies are most effective when applied in combination. For example, *TCRP Report 128: Effects of Transit-Oriented Development (TOD) on Housing, Parking, and Travel*, surveyed 17 housing projects that combined compact land use with transit access and found that these projects averaged 44 percent fewer vehicle trips per weekday than that estimated by the Institute for Transportation Engineers (ITE) manual for a typical housing development.[104] The *Moving Cooler* study also found that transit and nonmotorized improvements were more effective in areas of higher population density.[105] It further might be expected that strategies that encourage the use of alternative modes (such as road pricing) would have a greater impact when applied in conditions when better alternatives exist (as would be found with increased transit investment and more compact land use patterns), although evidence on the interactive effects among all of these factors in combination is limited.

- Research combined with pricing signals and or technology forcing regulations can reinforce one another. Federal research investments may be successful in developing new alternative fuels and fuel efficiency technologies, but without a market, these new technologies will not be introduced. For instance, the Partnership for a New Generation of Vehicles, begun in 1993 as a partnership between the Federal government and Detroit automakers, produced 60 to 80 mpg diesel hybrid prototypes, however, these new vehicles were never put into production. This illustrates the need not just for research but for incentives and long-term carbon price signals to spur mass production of low-carbon vehicles. Similarly, markets may exist for low-carbon vehicles but auto manufacturers may be reluctant to invest heavily in technology development because of the large investments needed and the high risks for product failures.

- Some pricing strategies may be redundant with each other, although higher prices through multiple mechanisms would of course have greater GHG reduction benefits, and some measures can be complementary. Among

---

[104] G.B. Arrington and Robert Cervero. *Transit Cooperative Research Program (TCRP Report 128: Effects of TOD on Housing, Parking and Travel*. Transportation Research Board: Washington, DC, 2008.

[105] Cambridge Systematics, Inc. (2009). *Moving Cooler: An Analysis of Transportation Strategies for Reducing Greenhouse Gas Emissions*. Urban Land Institute: Washington, D.C.

transportation-specific pricing mechanisms, raising the existing gas tax to levels higher than what cap and trade would cause entails almost no administrative costs and provides an incentive to purchase more efficient vehicles, but is politically unpopular. Other pricing mechanisms, such as a VMT fee or pay-as-you-drive (PAYD) insurance, would not encourage vehicle efficiency gains unless the VMT fees were differentiated by GHG emission rates or weights of different vehicles. PAYD insurance has the advantage of providing the majority of consumers with net cost savings. The technology for implementing either a VMT fee or PAYD insurance also could support congestion pricing, which would have the additional benefit of improving system efficiency and reducing travel times.

3.9 COBENEFITS

Strategies also can be compared according to their cobenefits. All of the strategies will result in lowered consumption of petroleum, and as such may have national security benefits to the extent that U.S. dependence on petroleum imports is reduced. Land use, transit, and nonmotorized strategies also will reduce household expenditures on fuel and on vehicle operating and ownership costs by reducing demand for carbon-intensive travel. Vehicle efficiency and system efficiency strategies will reduce household expenditures on fuel through more fuel efficient travel. Table 3.2 shows the estimated savings resulting from the system efficiency and travel activity strategies analyzed in this report. These can be compared with projected fuel use in 2030 for all transportation sources of 16.8 million barrels per day oil equivalent, or the equivalent of about 288 billion gallons of gasoline annually.[106] As with the greenhouse gas reduction benefits, fuel savings for the individual strategies or strategy families cannot be added together.

---

[106] AEO Reference case, April 2009 release, Table 7, and assuming 47 gallons of gasoline per barrel of oil equivalent.

Table 3.1   Potential Petroleum Savings in 2030[107]

Strategy Family		Petroleum Savings (billions of gallons of gasoline and diesel)	
		Low	High
Price Carbon		6.3	10.4
Improve Transportation System Efficiency	On-road	4.6	8.0
	Air, Rail, Marine	1.8	5.1
Reduce Carbon-Intensive Travel Activity		12.1	40.3

Source: Cambridge Systematics analysis.

Note: Vehicle efficiency and low carbon fuels strategies are not included here because of ongoing rulemakings.

NHTSA, in its preliminary rulemaking for revised CAFE standards as required by the Energy Independence and Security Act, reviews literature on the economic costs of dependence on foreign oil, and therefore the benefit of fuel savings resulting from increased CAFE standards. NHTSA estimates the benefits related to oil supply disruptions and monopsony costs (higher prices for petroleum products resulting from the effect of U.S. oil import demand on the world oil price) to range from about $0.108 to $0.539 per gallon saved, with a best estimate of $0.298 per gallon. These estimates do not include reduced outlays for military operations, as NHTSA concludes that fuel efficiency standards will not materially affect these costs.[108]

As vehicle fuel efficiency and low-carbon fuel strategies are implemented, transportation continues to fulfill the same function (moving people and goods) with little impact on mobility or accessibility. In contrast, most system efficiency strategies have significant mobility cobenefits, especially travel time savings and

---

[107] This table is based on rough estimates of fuel savings for individual strategies. For all strategies except low-carbon fuels, these estimates were derived by back-calculating fuel savings based on the GHG reductions from the strategy, considering the carbon content of gasoline, diesel, and/or jet fuel as appropriate for the strategy. Conversion factors of 9.16, 10.56, and 9.95 kg CO_2e per gallon were used respectively for gasoline, diesel, and jet fuel reflecting the carbon content of the fuel (8.81, 10.15 ,and 9.57 kg/gallon) inflated by 4 percent to account for non-CO_2 GHG emissions. The gasoline conversion factor was used for strategies affecting light-duty vehicle travel, the diesel factor for strategies affecting heavy-duty, rail, and marine travel, and the jet fuel factor for strategies affecting aviation. For strategies affecting all highway travel, factors were weighted 69 percent gasoline and 31 percent diesel, based on the fraction of these vehicles used in on-road vehicles as estimated from FHWA's Highway Statistics. .

[108] National Highway and Traffic Safety Administration (2009). *Corporate Average Fuel Economy for MY 2012-2016 Passenger Cars and Light Trucks: Preliminary Regulatory Impact Analysis.* Figures are in 2007 U.S. dollars.

resulting economic benefits from reduced congestion and travel times, whether by highway, transit, air, or rail. The primary exception is speed limit reduction, which reduces mobility and may increase shipping costs by increasing travel times. Land use and transit strategies reduce household transportation expenses and have mobility benefits for those who do not drive because of advanced age, young age, disability, or income. Finally, public health benefits can result from land use, nonmotorized, and transit strategies that encourage walking and biking.

For fuels strategies, the environmental and social impacts of biofuels production could be negative for those production pathways that require considerable amounts of land and compete with food supplies. Also, to the extent that fuels are produced domestically rather than from international sources, national security benefits may be achieved due to the reduced threat of energy supply disruption.

Travel activity strategies may have significant cobenefits or disbenefits. The most significant benefits result from improved mobility from improvements to alternative modes, including transit, ridesharing, and nonmotorized travel, as well as more compact land use patterns that support these alternatives. There can also be opposition to increased densities at the local level. The most significant disbenefits include mobility and equity impacts to lower-income populations from pricing strategies that increase the cost of carbon-intensive travel beyond their willingness or ability to pay without compensating increases in availability of less carbon-intensive, more affordable travel ammenities or other compensation mechanism. Pricing also faces substantial barriers in the form of public opposition and concerns over equity impacts, which may be addressed through redistribution of revenue and/or investment in alternative modes.

Many strategies reduce air pollution, but the reductions would vary depending upon the specific strategy. Wind and weather patterns also complicate the impacts. Reductions in total vehicle activity would reduce air pollutant emissions. More efficient vehicle operations (reduced idling, congestion, etc.) would further reduce air pollutant emissions beyond the levels from vehicle activity reduction, though NOx emissions would likely increase with speeds above 40-45 mph. Heavy-duty and off-road vehicles tend to have less strict emission controls than light-duty vehicles, however, so some strategies that reduce GHG emissions through switching travel or goods movement to more efficient modes (transit, freight rail, marine) may not reduce emissions of all pollutants, and may even increase some emissions (although Federal standards for heavy-duty vehicles and locomotives are leading to substantial improvements in these sectors). In some cases, localized benefits may be far more significant than the total quantity of pollutants reduced in a region.

Vehicle fuel efficiency strategies may reduce air pollutant emissions by reducing the amount of fuel burned. However, emission standards would be the primary factor influencing emissions levels. Some technologies (such as hybrid-electric powertrains) may make it easier to meet advanced emissions standards. The impacts of low-carbon fuel strategies on vehicle-based emissions may again be limited by standards, although some types of fuels may decrease or increase particular types of pollutants. Low-carbon fuels may also have significant emissions (or provide reductions) associated with their manufacture and transport. Table 3.3 shows life-cycle emissions for various alternative fuels compared to gasoline or diesel, considering *current* vehicle emissions control and air pollution control technology.

Table 3.2   Relative Life-Cycle Emissions of Alternative Fuels (Percent Change versus Conventional Gasoline)

Pollutant	Conventional Gasoline Emissions (g/mi)	CNG	LPG	Gaseous Hydrogen[b]	Battery Electric Vehicle[c]
VOC	0.316	-45%	-35%	-92%	-91%
CO	3.817	0%	0%	-98%	-98%
NO_x	0.379	-20%	-14%	-59%	-11%
PM_{10}	0.083	-9%	-47%	23%	416%
$PM_{2.5}$	0.036	-20%	-38%	36%	220%

Source: GREET Model Version 1.8b, with default assumptions for current vehicle technologies. Relative emissions will vary depending upon vehicle emission controls as well as fuel extraction and production methods. Relative emissions may change in future years as these various technologies evolve in different ways.

[a] Compared with diesel.

[b] Assuming distributed natural gas reforming.

[c] Assuming current grid-average electricity generation mix. Future scenarios will differ considerably depending upon grid mix and when vehicles are charged.

Emissions impacts from vehicles that are powered by electricity (including battery-electric vehicles, plug-in hybrid electric vehicles or PHEVs, and hydrogen fuel cell vehicles) cannot be compared on an apples-to-apples basis with emissions from internal combustion engine vehicles, as emissions will occur at different locations (away from the vehicle) and therefore have different air quality and health impacts. Battery-electric and hydrogen fuel-cell vehicles will result in zero emissions from the vehicle itself (as will PHEVs operating in all-electric mode), although total emissions from powerplants will increase slightly. Furthermore, using grid-average emissions for battery-electric vehicles may be inappropriate as emissions will depend upon when vehicles are charged (and will vary by region of the country).

3.10 INFRASTRUCTURE FINANCE

The Federal Highway Trust Fund was established in 1957 as a dedicated, user-funded source of revenue to fund the Interstate Highway System as well as other

Federal transportation programs. It is the primary source of revenue for most Federal surface transportation programs including the Federal-aid Highway Program and the Federal Transit Program. The Highway Trust Fund is funded primarily through taxes on motor fuels, as well as through excise taxes on truck tires, retail sales of heavy-duty vehicles and trailers, and other motor vehicle-related items. Fuel tax receipts made up 88 percent of Trust Fund revenue in FY 2008.[109] Gasoline is taxed at a rate of 18.4 cents per gallon and diesel at a rate of 24.4 cents per gallon. Taxes on alternative fuels, including liquefied petroleum gas and compressed and liquefied natural gas, are set equal to gasoline taxes on an energy-equivalent basis.[110] The vast majority of these taxes are deposited in the Highway Account, with 2.86 cents per gallon (15.5%) directed to the Mass Transit Account and 0.1 cents to the Leaking Underground Storage Tank Trust Fund. Net receipts in FY 2007 were $34.3 billion to the Highway Account and $5.0 billion to the Mass Transit Account.[111] States also fund highway and other transportation programs through motor fuel taxes, although tax rates vary by state.

In 2009, the Highway Trust Fund was projected to go into a negative balance, with cumulative outlays exceeding cumulative income, and a cash shortfall was averted by means of a $7 billion cash transfer from the General Fund.[112] This was in addition to a previous cash shortfall in 2008 which was averted through a transfer in that year of $8 billion from the General Fund.[113] This situation indicates a lack of revenue raised from users of the transportation system compared to current levels of Federal expenditure. Therefore, the GHG reduction strategies described in this report that would entail Federal funding may require reprioritization of current expenditures or additional taxation in order to implement.

Because fuel taxes are collected by the gallon, funding challenges may be compounded in the future since most of the on-road vehicle strategies analyzed in this report reduce total motor vehicle fuel use, and therefore (unless the tax were modified to be a sales-based tax) would reduce total Federal Highway Trust Fund revenues (as well as State fuel tax revenues) in rough proportion to fuel savings (and related GHG reductions). For example, if advanced light-duty gasoline vehicles were to achieve a 20 percent efficiency improvement by 2030 and reach a market penetration of 60 percent, Highway Trust Fund revenues

---

[109] U.S. DOT, Office of the Inspector General (2009). "Highway Trust Fund Solvency." Testimony to Senator Judd Gregg, June 24, 2009.

[110] Federal Highway Administration (2008). *Highway Statistics 2007*, Table FE-21B.

[111] Federal Highway Administration (2008). *Highway Statistics 2007*, Table FE-10.

[112] The Legislative Services Group, *Transportation Weekly*, Volume 10, Issue 34, August 3, 2009.

[113] U.S. DOT, 2009 (cited).

would decline by 8 percent, or about \$3.1 billion compared to FY 2007 receipts.[114] Strategies focused on heavy-duty vehicles would have a somewhat greater impact than those focused on light-duty vehicles because of the higher tax rate on diesel fuel. A shift from gasoline to diesel light-duty vehicles would have a smaller revenue impact; while less total fuel is consumed, the tax rate on diesel fuel is higher.

The revenue effects of alternative fuels will depend upon taxation policy. Since 2006, Federal policy has been to tax alternative fuels on an energy-equivalent basis to gasoline.[115] This means that policies focused solely on increasing the use of lower-carbon fuels should not have a significant revenue impact on the Highway Trust Fund. On the other hand, current tax policies may have implications for general fund revenue; for example, ethanol receives a substantial tax credit which reduces general fund revenues, but a tariff is levied on imported ethanol which could potentially generate revenue. The long-term finance implications of a shift to hydrogen or electricity will depend on tax policy for these fuels; current Federal policy does not tax these fuels for transportation purposes, and therefore any shift to these fuels would result in lost revenues to the Highway Trust Fund under the current finance structure.

Pricing measures, such as a cap and trade system, carbon tax, VMT fee, or congestion pricing, would provide a new or alternative revenue source, which could potentially be directed to transportation infrastructure finance. (The revenue impacts of a cap and trade system would depend upon the extent to which allowances are auctioned vs. given away.) While transportation-specific pricing revenues are likely to be redirected towards transportation system investment, it is less certain that a portion of revenues from economy-wide measures such as cap and trade allowances or carbon taxes would be redirected towards transportation. Finally, pricing measures that reduce the wear and tear on existing road networks could also have fiscal implications to the extent they lower the total cost of maintaining or improving system performance even if investments in alternatives to carbon-intensive travel are increased.

---

[114] Gasoline tax receipts accounted for about \$25.5 billion of trust fund revenue in FY 2007, or about two-thirds of total receipts, so 20 percent * 60 percent * 2/3 = an 8 percent reduction or \$3.1 billion. This calculation assumes that trust fund revenues remain constant through 2030, which is consistent with AEO Reference case projections of relatively constant fuel consumption (with increases in VMT offsetting increases in fuel efficiency).

[115] Ethanol users actually pay slightly more per gallon of gasoline equivalent even though ethanol is taxed less on a volumetric basis (Vol. 2 Sec. 2.2).

3.11 SUMMARY OF FINDINGS

Tables 3.5 – 3.8 present a consolidated overview of GHG reduction strategies, summarizing a wide range of specific information. The table includes the following information for most strategies:

- **Key Deployment Assumptions** – Key assumptions about the strategy that affect the magnitude of results.

- **Effectiveness:**

  - **Percent GHG Reduction** – Percent reduction in GHG emissions from baseline, for:
    - ° **Transportation Sector** – Reduction as a percentage of total transportation sector baseline emissions (based on Annual Energy Outlook March 2009 Reference case) in 2030 (2,171 million metric tons carbon dioxide equivalent, or mmt CO_2e).
    - ° **Relevant Subsector(s)** – Relevant transportation subsector(s) – light-duty vehicle (LDV), heavy-duty vehicle (HDV), and on-road vehicles, rail, marine, and aircraft – that strategy affects and percentage reductions for this subsector.

  - **Absolute GHG Reduction** – Absolute reduction in year 2030 or 2050, expressed in million metric tons CO_2e; range of values (lower/upper) indicated when findings differ. Values for 2050 are shown only if significantly different than for 2030. "N/A" signifies that values may be significantly different in 2050 than 2030, but were not modeled in this timeframe.

  - **Timing of Benefits** – If the strategy is implemented today, this is a projection of how long it would take to achieve the reductions noted. Three ranges: i) Short – most benefits can be achieved within five years; (ii) Mid – most benefits achieved within 5 to 20 years; (iii) Long – most benefits not achieved for at least 20 years.

- **Cost Effectiveness** – Expressed in \$/metric ton CO_2e; range of values (lower/upper) indicated when findings differ. The cost-effectiveness estimates should be read with caution because they reflect monetary costs only. They do not reflect other very significant benefits or disbenefits to consumers such as travel time impacts, utility of foregone trips, health benefits, air quality impacts, and increased or decreased accessibility or mobility. Taxes, fees, and rebates are not included in cost-effectiveness calculations, since they are regarded as a transfer payment (from the private sector to the public sector). However, the imposition of taxes, fees, and rebates may create welfare changes that are difficult to monetize but nonetheless represent a real cost or benefit to consumers. The two types of cost effectiveness cited are:

- **Direct Implementation Costs**—this accounts only for costs required to implement the strategy, such as the cost to transit agencies to provide increased public transportation services or the cost to State highway departments to time traffic signals. It includes infrastructure construction costs, capital costs, ongoing maintenance and operations costs, program administrative costs, etc. It does not include any monetary savings such as decreased fuel or vehicle operating costs.

- **Net Included Costs**—this includes direct implementation costs (such as the increased cost of high technology vehicles over conventional vehicles or the cost of telecommuting equipment) as well as monetary savings such as savings from reduced fuel use and reduced vehicle operating costs. Costs included may vary by source (see Appendix A).

Both direct implementation costs and net included costs are provided for system efficiency and travel activity strategies because for these strategies, the costs are primarily borne by the public sector and the savings primarily accrue to individuals. Only net included costs are provided for vehicle and fuel strategies because both costs and savings typically accrue to the same entity—the vehicle owner.

With some exceptions, costs in this report are expressed in present-year real dollars (as cited in the data source or reference) without any inflation or discounting. In a few cases, when cost estimates were particularly old (e.g., prior to year 2000), the consumer price index was applied to inflate values to current year dollars. When calculating cost effectiveness, future-year operating cost savings for on-road vehicles (but not for off-road vehicles) were discounted using a discount rate of seven percent. The cost-effectiveness estimates computed from the *Moving Cooler* study data are also based on discounting future vehicle operating cost savings at a rate of seven percent. Cost-effectiveness estimates from other studies cited in this report that included future cost savings may have used other discounting assumptions.

Table 3.3 Findings by Strategy: Carbon Pricing and Low-Carbon Fuels

Strategy	Key Deployment Assumptions	Effectiveness							Timing of Benefits	Cost Effectiveness ($/tonne CO_2e)
		Percent GHG Reduction					Absolute GHG Reduction (mmt CO_2e/year)			
		Transportation Sector		Relevant Subsector(s)						
		2030	2050	Subsector	2030	2050	2030	2050		Net Included Costs
Economy-Wide Market-Based Strategies										
Cap-and-Trade / Carbon Tax	Allowance price or tax of $30 to $50/ton CO2e in 2030	2.6-8.5%	n/a	All	2.6-8.5%	n/a	53-174	n/a	Short-Long	n/a
Motor Fuel Taxes	Equivalent to ~$0.20 to $2.40/gal	2.4-23%	n/a	On-road vehicles	3.2-32%	n/a	50-500	n/a	Short-Long	n/a
Low-Carbon Fuels										
Ethanol					Not analyzed					
Biodiesel					Not analyzed					
Natural gas	2.5-5% of total U.S. natural gas use diverted to transportation; 15% GHG reduction per vehicle	0.3-0.6%		LDV	0.6-1.2%		7-13		Short	($130)-($50)
Liquefied petroleum gas (LPG)	Two times current consumption rates	0.01%		LDV	0.02-0.03%		0.2-0.3		Short	n/a
Synthetic fuels	Not analyzed								Mid-Long	
Hydrogen	2030 – 18% LDV market penetration, 40-55% GHG reduction per vehicle; 2050 – 60% LDV market penetration, 79-84% GHG reduction per vehicle	2.4-3.4%	18-22%	LDV	4.8-6.8%	36-44%	52-74	390-470	Long	($194)-$275
Electricity	2030 – 5% LDV market penetration, 68-80% GHG reduction per vehicle; 2050 – 56% LDV market penetration, 78-87% GHG reduction per vehicle	2.2-2.5%	26-30%	LDV	4.3-5.1%	53-59%	47-55	570-640	Long	($90)-$343
Aviation Fuels	Not analyzed									

Table 3.4   Findings by Strategy: Vehicle Fuel Efficiency

Only per vehicle GHG reductions are provided for vehicle efficiency strategies. Percent GHG reduction for the transportation sector as a whole from each strategy will be much less than the per vehicle reductions since each vehicle type comprises only a portion of the total transportation sector, market penetration will almost certainly be less than 100 percent, and fleet turnover time will delay realization of benefits.

Strategy	Per Vehicle GHG Reduction Compared to Conventional Vehicle, 2030
On-Road Light-Duty Vehicles	
Advanced Conventional Gasoline Vehicles	8 – 30%
Diesel Vehicles	16%
Hybrid Electric Vehicles	26 – 54%
Plug-In Hybrid Electric Vehicles	46 – 75%
On-Road Heavy-Duty Vehicles	
Retrofits of heavy-duty trucks to use aerodynamic fairings, trailer side skirts, low-rolling resistance tires, aluminum wheels, and planar boat tails	10 – 15%
Powertrain and Resistance Reduction for New Trucks	10 – 30%
Transit hybrid electric buses	10 – 50%
Rail	
Power System Modifications	
• Common rail injection systems	5 – 15%
• Genset engines	35 – 50%
• Hybrid yard engines	35 – 57%
• Hybrid line-haul operations	10 – 15%
Train Efficiency Improvements	
• Light weight railcars, aerodynamics, wheel to rail lubrication	4-10% individually
• Improving load configuration for intermodal trains	up to 27%
Marine	
Improvements to Ship Design and Propulsion Systems	4 – 15% for ship design
	Up to 20% for diesel electric for vessels that change speed or load frequently (cruise ships, harbor tugs, and ferries)
Aircraft	
Engine technology and airframe improvements	1.4-2.3%[*]
Vehicle Air Conditioning Systems	**Reduction in Mobile Air Conditioner GHGs**
Can-Ban (Ban on do-it-yourself air conditioner servicing)	66% (California study)
Alternative Refrigerant Chemicals	91.3 to 99.9% depending on refrigerant type and mechanical efficiency

[*]Fleet-wide annual aircraft efficiency improvement during 2015-2035 relative to 2015 as the base year.

Table 3.5 Findings by Strategy: System Efficiency

Strategy Name	Key Deployment Assumptions	Effectiveness					Cost Effectiveness ($/tonne CO₂e)	
		Percent Reduction in 2030			Absolute GHG Reduction in 2030 (mmt CO₂e/year)	Timing of Benefits	Direct Implementation Costs Only	Net Included Costs[a]
		Transportation Sector	Relevant Subsector(s)	Reduction for Subsector				
System Efficiency								
Highway Operations and Management								
Traffic Management	Deployment of full range of traffic management strategies on freeways and arterials at rate of 700 to 1,400 miles/year	n/a	On-road vehicles	n/a	n/a	Short-Mid	$40->$2,000	($120)->$2,000
Real-Time Traveler Information	Deployment of highway traffic information at same rate as traffic management	n/a	On-road vehicles	n/a	n/a	Short-Mid	$160->$600	0->$500
Highway Bottleneck Relief	Improve top 100 to 200 bottlenecks by 2030	n/a	On-road vehicles	n/a	n/a	Mid-Long	n/a	n/a
Reduced Speed Limits	55 mph national speed limit	1.2-2.0%	On-road vehicles	1.7-2.7%	27-43	Short	$10	($320)
Truck Operations and Management								
Truck Idling Reduction	26-100% of sleeper cabs with on-board idle reduction technology	0.1-0.3%	HDV	0.4-1.2%	2-6	Short-Mid	$20	($420)-($480)
Truck Size and Weight Limits	Allow heavy/long trucks for drayage and noninterstate natural resources hauls	0.03%	HDV	<0.1%	0.6	Short	$0	($1,200)
Urban Consolidation Centers	Large/high density urban areas	0.01%	HDV	<0.1%	0.2-0.3	Mid	$30-60	($300)
Freight Rail and Marine Operations								
Freight Modal Diversion	Rail infrastructure improvements-up to 6% avoided diversion of rail traffic to truck	0.0-0.2%	HDV + rail	0.0-0.8%	0.2-5	Mid	$80-200	n/a
Rail and Intermodal Terminal Operations	Not analyzed							
Ports and Marine Operations	Land and marine-side operational improvements at container ports	0.01-0.02%		n/a	0.2-0.4	Short-Mid	n/a	n/a
Air Traffic Operations								
Air Traffic Operations	Air traffic management in U.S. airspace	0.3-0.7%	Domestic Aircraft	2.5-6% (cum. thru 2035)	8.9-25.2	Mid	n/a	<$0
Infrastructure Construction and Maintenance								
Construction Materials	Fly-ash cement and warm-mix asphalt used in highway construction throughout U.S.	0.7-0.8%	n/a	n/a	15-18	Short	$0-$770	$0-$770
Other Transportation Agency Activities	Alternative fuel DOT fleet vehicles, LEED-certified DOT buildings	0.1%	n/a	n/a	2-3	Mid	n/a	n/a

[a] "Net included costs" typically includes implementation costs and vehicle operating cost savings, but not travel time costs/savings or other non-monetary costs and benefits.

Table 3.6 Findings by Strategy: Reduce Carbon-Intensive Travel Activity

Strategy Name	Key Deployment Assumptions	Transportation Sector	Relevant Subsector(s)	Reduction for Subsector	Absolute GHG Reduction in 2030 (mmt CO₂e/year)	Timing of Benefits	Direct Implementation Costs Only	Net Included Costs[a]
		Effectiveness					Cost Effectiveness ($/tonne CO₂e)	
		Percent Reduction in 2030						
Travel Activity								
Pricing								
VMT Fees	VMT fee of 2 to 5 cents per mile	0.8-2.3%	On-road vehicles	1.1-3.1%	17-50	Short	$20-$280	($650)-($910)
Intercity Tolls	Toll of 2 to 5 cents per mile on rural Interstate highways	0.1%	On-road vehicles	0.1-0.2%	1-3	Short	$500-$800	$50-($630)
Pay-as-You-Drive Insurance	Require states to permit PAYD insurance (low)/Require companies to offer (high)	1.1-3.5%	LDV	1.4-4.7%	23-75	Short	$30-$90	($960)
Congestion Pricing	Maintain level of service D on all roads (average fee of 65 cents/mile applied to 29 percent of urban and 7 percent of rural VMT)	0.4-1.6%	On-road vehicles	0.6-2.2%	19-43	Short	$300-$500	($440)-($570)
Cordon Pricing	Cordon charge on all U.S. metro area CBDs (average fee of 65 cents/mile)	0.1%	On-road vehicles	0.1-0.2%	2-3	Short	$500-$700	($530)-($640)
Alternative Modes								
Transit Expansion, Promotion, and Service Improvements	2.4-4.6% annual increase in service; increased load factors	0.3-0.8% (2030) 0.4-1.5% (2050)	LDV	0.6-1.7% (2030) 0.8-3.0% (2050)	6-18 (2030) 9-32 (2050)	Mid	$1,200-$3,000	($900)-$1,000
Intercity Passenger Bus and Rail	Intercity rail 20 percent higher ridership/service increase than baseline; 11 new HSR corridors; Intercity bus 3 percent annual increase	0.0-0.3%	LDV	0.1-0.6%	1-6	Mid	$400-$1,400	($600)-$1,000
Non-motorized Transportation	Comprehensive urban pedestrian and bicycle improvements implemented 2010 to 2025	0.2-0.6%	LDV	0.4-1.1%	4-12	Mid	$80-$210	($600)-($700)
Land Use and Parking								
Land Use	60-90% of new urban growth in compact, walkable neighborhoods (4,000+ persons/sq mi or 5+ gross units/acre)	1.2-3.9% (2030) 2.5-7.7% (2050)	LDV	2.5-7.8% (2030) 5.0-16% (2050)	27-84 (2030) 56-170 (2050)	Long	$10	($700)-($800)
Parking Management	All downtown workers pay for parking ($5/day average for those not already paying)	0.2%	LDV	0.3%	3-4	Mid	n/a	n/a
Commute Travel Reduction								
Demand Management/	Widespread employer outreach and alternative mode support	0.1-0.6%	LDV	0.2-1.1%	6-14	Short	$30-$180	($1,000)

Strategy Name	Key Deployment Assumptions	Effectiveness					Cost Effectiveness ($/tonne CO2e)	
		Percent Reduction in 2030			Absolute GHG Reduction in 2030 (mmt CO2e/year)	Timing of Benefits	Direct Implementation Costs Only	Net Included Costs[a]
		Transportation Sector	Relevant Subsector(s)	Reduction for Subsector				
Travel Activity								
Commuter Measures								
Commute Travel Reduction (continued)								
Teleworking	Doubling of current levels	0.5-0.6%	LDV	0.9-1.2%	10-13	Short	$1,200-$2,300	$180
Compressed Work Weeks	Minimum—75% of government employees; Maximum—double current private participation	0.1-0.3%	LDV	0.3-0.6%	3-7	Short	n/a	n/a
Flexible Work Schedules	Not analyzed							
Ridematching, Carpool, and Vanpool	Extensive rideshare outreach and support	0.0-0.2%	LDV	0.1-0.5%	1-5	Short	$80	n/a
Public Information Campaigns								
Marketing Campaigns	Mass marketing in 50 largest urban areas; Individualized marketing reaching 10 percent of population	0.3-0.4%	LDV	0.5-0.8%	6-8	Short	$90-$270	n/a
Information on Vehicle Purchase	Expansion of SmartWay and other information campaigns	0.1-0.2%	On-road vehicles	0.2-0.5%	2-5	Short-Mid	n/a	n/a
Driver Education/ Eco-Driving	Minimum—Reach 10% of population; Maximum—Full penetration	0.8-4.3%	On-road vehicles	1.1-5.9%	18-94	Short-Mid	n/a	$0-($230)

[a] "Net included costs" typically includes implementation costs and vehicle operating cost savings, but not travel time costs/savings or other non-monetary costs and benefits.

4.0 Cross-Cutting Strategies

Two general strategies that cut across other strategy groups are addressed in this section. These are:

- **Transportation planning** and investment efforts, which can improve the operating efficiency of the multimodal transportation network and integrate transportation and land use planning to reduce travel distances; and

- **Pricing carbon** through a cap and trade system, carbon tax or increased motor fuels tax (which would affect vehicle fuel efficiency), encourage use of low-carbon fuels, and encourage more energy-efficient travel patterns.

4.1 TRANSPORTATION PLANNING AND INVESTMENT

The level of GHG emissions from transportation depends on the carbon content of the fuels, the fuel efficiency of the vehicles, the efficiency of the transportation system, and the level of travel activity. These latter two factors — the efficiency of the system and the level of travel activity — can be directly influenced through decisions that are made by Federal, State, regional, and local governments regarding the planning, funding, design, construction, and operations of the Nation's transportation systems.

Coordinating transportation and land-use decisions and investments enhances the effectiveness of both and increases the efficiency of Federal transportation spending. In most communities, jobs, homes, and other destinations are located far away from one another, necessitating a separate car ride for every errand and long delivery routes for goods. Strategies that support mixed-use development, mixed-income communities, and multiple transportation options help to reduce traffic congestion, lower transportation costs, improve access to jobs and opportunities, and reduce dependence on foreign oil, in addition to reducing greenhouse gas emissions. Prioritizing through planning low carbon alternatives such as public transportation, pedestrian facilities for biking, and walking, and lower carbon freight options such as rail or marine, can reduce GHGs, especially when deployed with synergistic policies such as land use. Similarly, prioritizing strategies such as signal timing, real-time traveler information, faster clearance of incidents, congestion pricing, freeway ramp meeting, and other intelligent transportation systems can reduce the pressure for new capacity while modestly reducing GHG emissions.

The Federal government is an important partner with State and local governments in shaping the Nation's surface transportation infrastructure. The

Federal government currently provides $52 billion[116] in funding for surface transportation annually, and Federal statute and regulations establish requirements for States and metropolitan planning organizations (MPOs) to undertake planning to determine how to use these resources. The Federal government also influences the efficiency of the Nation's air transportation system by operating the air traffic control system and providing assistance to improve the capacity and safety of airports, and provides funding for investments in rail and marine modes as well. Federal leadership on GHG mitigation and climate change planning can help convey the importance of GHG reduction to State and local transportation agencies. Furthermore, Federal coordination of housing, transportation, and environmental policies is key. A lack of coordination between these policies has contributed to the growth in vehicle miles traveled and GHG emissions.

Before discussing in more detail integrating climate change considerations into a transportation planning process, it is important to place this integration into the broader context of the current planning process. Planning is the information-based policy framework by which communities prepare and follow a reasoned course of action to achieving a desirable future vision. Plans represent blueprints for communities to follow, enabling them to evolve in an optimal way and influencing urban and rural development, economic prosperity, environmental quality, and social equity. Planning is a cooperative process, bringing together a wide range of perspectives from different people, organizations, and stakeholder groups to pursue common ground on a variety of issues. As such, it must consider a wide range of forces—such as mobility, health, economic growth, environmental sustainability, and land use—in determining a community's ideal vision and identifying the priority projects, programs, and strategies for achieving that vision. The transportation system, and its GHG impact, is one element among many societal concerns.

Planning includes comprehensive consideration and choice of preferred action from a range of possible strategies. Successful planning depends upon an information-driven evaluation process that encompasses diverse viewpoints, the collaborative participation of relevant agencies and organizations, and open, timely, and meaningful public involvement. Without broad and meaningful participation, there is a risk of making poor decisions, or decisions that have unintended negative consequences. On the other hand, having broad participation makes it possible for all parties to work together in partnership to make a lasting contribution to an area's quality of life. The public includes anyone who resides, has an interest in, or does business in a given area potentially affected by the decisions, as well as regional and national representatives. Federal, State, and local agencies with an interest in the region

---

[116] Not including additional 2009 transportation funding from the American Recovery and Reinvestment Act of 2009.

play a particularly important role in the achieving the vision. Many of those agencies have statutory responsibilities that impact planning decisions. Coordination and cooperation among all interested parties and relevant agencies is necessary to achieve the vision. This is particularly important as State and local transportation planners do not often have control over land use decisions, but can serve as conveners of stakeholder groups and work closely with land use planning authorities. Similarly, by providing funding and requiring a planning process, the Federal government is an important stakeholder, but much decision-making power appropriately resides at State and local levels.

While planning is an open and collaborative process, it also is disciplined by the need to abide by important fiscal and environmental constraints. These constraints limit the extent of projects and strategies that may be recommended in plans, forcing communities to make difficult tradeoffs. In the end, a plan represents the community's preferred actions, limited to those that are achievable within reasonable constraints.

Federal statute requires that States and MPOs engage in a transportation planning process and develop a plan that "include[s] both long-range and short-range program strategies/actions that lead to the development of an integrated intermodal transportation system which facilitates the efficient movement of people and goods." They must also develop a short-range program of transportation improvements, based on the long-range transportation plan, designed to achieve the area's goals using spending, regulating, operating, management, and financial tools. Transportation agencies confront a wide range of tradeoff decisions within and between modes, policy objectives, performance goals, geographic regions, and market segments when developing these plans and programs. Therefore, any decision on GHG reduction activities, including where to invest limited resources, needs to be balanced with its impact on other goals and priorities.

There are three main ways in which the Federal government can influence GHG reduction through transportation infrastructure planning and investment: technical assistance, regulations, and incentives. When considering each of these avenues, the Federal government can learn from the experiences of States and MPOs on incorporating climate change considerations into their transportation planning processes. As documented in a recent study,[117] these experiences vary widely. Climate change can appear in the vision, goals, policies, strategies, trends, challenges, and performance measures of long-range transportation plans. Some plans merely recognize that climate change is an issue that relates to transportation and begin to point out the relevance of existing plans and strategies to climate change. Other plans make climate change more central to

---

[117] ICF International (2008). *Integrating Climate Change into the Transportation Planning Process*. Prepared for Federal Highway Administration.

their goals and policies. Some include innovative analyses of the GHG impacts of various alternatives that could serve as models for other areas.[118]

There are a range of options for the Federal government to work with State and local governments to address climate change. The range of options include providing technical assistance, including climate change as a planning factor, providing funding incentives, requiring states and MPOs to develop strategies for reducing transportation GHGs, establishing mandatory GHG reduction targets, and aligning Federal funding distribution with performance measures. Each option will have differing levels of impact on GHG emissions and on the level of effort required.

Technical Assistance

The DOT provides technical assistance to States and metropolitan areas to support transportation planning and could provide increased technical assistance on climate change issues. Technical assistance also can be provided to other public and private sector entities responsible for transportation infrastructure and services, such as port authorities, airports, and railroads. Recent DOT technical assistance on climate change issues includes workshops around the country with State DOTs and MPOs, release of a report on integrating climate change considerations into the transportation planning process, assistance on transit-oriented development planning, and analysis of State climate action plans. The DOT will continue to provide such assistance.

Technical assistance can be provided on a variety of topics, such as:

- Procedures for developing greenhouse gas inventories and analyzing and evaluating the benefits of alternative plans and projects with respect to GHG emissions (EPA's MOVES model provides an example);

- Data collection and model techniques critical to GHG emissions analysis. Examples include supporting the development of robust GHG sketch-planning tools, supporting the broader application of integrated transportation and land use models, and restoring or expanding funding for critical data collection efforts such as the Vehicle Inventory and Use Survey;[119] and

- Scenario planning, visioning and integrated transportation and land use planning. Examples include guidelines and best practices for planning processes, implementation practices such as model ordinances, assistance with zoning code updates, and assistance with updating roadway design standards to accommodate multimodal travel.

---

[118] ICF International, 2008 (cited).

[119] The Vehicle Inventory and Use Survey served as the primary source of information on energy-related characteristics of the nation's vehicle fleet from 1963 through 2002.

A recent Federal Highway Administration study found that quantification of GHG emissions is one of the most challenging aspects of integrating climate change into transportation planning for States and MPOs. There is room for improvement across the board in inventory techniques and techniques for estimating the impact of policies and strategies.[120] Technical assistance also can provide States, MPOs, and other planning entities with greater ability to work with stakeholders and the public to select the most effective and cost-effective GHG reduction strategies.

Technical assistance can be accommodated within the existing legislative and regulatory process. Simply improving planning and analysis capabilities, however, is no guarantee that GHG reduction strategies will be implemented. In addition, technical assistance would be of limited value to local agencies if they do not have the resources to implement or make use of improved methods. The benefits of technical assistance would be increased if it is accompanied by funding to support the use of specific planning or analytical methods.

Regulations

Federal regulations that direct State and metropolitan transportation planning can influence GHG reductions through a number of avenues. Changing these regulations may require legislative authorization, followed by a DOT rulemaking process to revise the existing regulations. Other actions to change planning could be done under current law. In addition, regulations can be established (through legislative authorization) that expand the scope of Federal influence in transportation planning to other modes or geographic scales.

GHG Consideration

The most direct approach would be to establish explicit requirements to consider GHG within the statewide and/or metropolitan planning process.[121] The current Federal transportation planning statutes and regulations include a number of requirements that generally align with climate change mitigation, such as requiring that plans "protect and enhance the environment [and] promote energy conservation" and that plans discuss "potential environmental mitigation activities."

Perhaps the mildest form of additional regulation would be to require consideration of greenhouse gas emissions as a transportation planning factor, without including any specific requirements as to how GHG should be addressed. While climate change and GHG reduction already can be considered under the energy and environment planning factor, creating a stand-alone provision would reinforce the importance of this issue within the planning

---

[120] ICF International, 2008 (cited).

[121] Highway and transit transportation planning requirements are found in 23 USC 134, 23 USC 135, 49 USC 5303, and 49 USC 5304, as well as 23 CFR 450.

process. More importantly, many states and MPOs currently use planning factors to define their State or regional transportation goals – general statements of purpose that reflect a long-term desired end to a specific transportation need or issue. A goal is typically very qualitative in nature, and is often only used to communicate broader investment strategies to the public. However, when used as part of a performance-based planning framework, goals are the key first step in identifying potential solutions that address specific transportation needs, and indicate a general direction for transportation investment. This becomes the foundation for establishing performance measures which provide a mechanism to "test" solutions and provide a quantitative means to describe the impact of a project (or group of projects). Some States and MPOs are moving towards performance-based planning (also an emphasis area in reauthorization) and could use a GHG planning factor to support project and/or systems-level evaluation of GHG reduction to demonstrate consideration of the factor.

More prescriptive actions would include a requirement to consider GHG mitigation measures in plan development, or a requirement to develop GHG inventories and forecasts for plan alternatives. This action would require DOTs and MPOs to develop 1) a baseline inventory of existing GHG emissions from transportation sources in their State or region, and 2) GHG forecasts associated with each alternative evaluated in the long-range transportation plan. For some MPOs the requirement could be relatively straightforward, as the MPO already prepares baseline and plan alternative model runs using its regional travel demand model, and a first-level GHG analysis could be added without much effort. However, data and modeling improvements might be needed in many metropolitan areas to develop better GHG estimates from strategies such as traffic operations, transit, nonmotorized, land use design, and freight intermodal improvements, which many of today's models are not designed to analyze. The requirements would have more significant implications for State DOTs, which typically do not develop a full network model with a comprehensive set of statewide projects to analyze plan alternatives. The inclusion of nonhighway modes in the inventory requirement also would add another level of data collection and analysis that does not currently exist at either level.

Such requirements would ensure that planning agencies consider the GHG impacts of their actions. Inventory and forecast requirements would ensure that information is introduced to inform decision-making. As with technical assistance, however, these requirements would not guarantee that GHG-reducing strategies can be implemented. The acceptability of such requirements to transportation planning agencies would likely decline in relationship to their specificity. A statewide requirement to develop an inventory and forecast would require the development of new technical capabilities and planning activities by State DOTs. Resource requirements also would depend upon the type of analysis required. A VMT-based inventory is relatively simple, but a more detailed and precise inventory that captures factors such as vehicle operating conditions is not supported by current tools.

Because of the global nature of climate change, GHG emissions impacts and reduction strategies are more effectively evaluated and addressed at a regional or systems level than at the transportation project level. A voluntary regional-level analysis of transportation-related GHG emissions and reduction strategies analysis may be appropriate at the planning stage. Some States already have requirements to conduct GHG analysis for projects subject to State environmental review requirements. Where such requirements are in place, or where voluntary regional analyses are conducted, National Environmental Policy Act (NEPA) documents could summarize information regarding regional-level analysis of transportation related GHG emissions and reduction strategies from transportation plans and associated studies. The Council on Environmental Quality is developing guidance on consideration of climate change in NEPA documents. Any DOT guidance on NEPA would need to be consistent with Council on Environmental Quality guidance.

Integrated Land Use and Scenario Planning

Better integrating transportation and land use planning is a major strategy governments can undertake to improve access to housing, jobs, and other destinations while reducing travel distances, and consequently GHG emissions. Transportation and land use are interdependent. Decisions on the locations and densities of housing, retail, offices, and commercial properties impact travel patterns to these destinations. Similarly, the geographic placement of roads, public transportation, airports, and rail lines influences where homes and businesses are built. However, transportation planning and land use planning often occur separately, frequently resulting in longer travel distances and higher GHG emissions. By determining where and what type of transportation infrastructure is built, and thus the travel options available, State and metropolitan transportation planning greatly influences travel patterns, land use, energy consumption, and, as a result, GHG emissions. Integrated transportation and land use planning practices that promote clustered or higher density, mixed use development, and colocation of services near transit can reduce emissions by shortening driving distances. Similarly, infill, connected street networks, traffic calming, sidewalks, bike lanes, and walking paths can provide alternatives to carbon intensive travel.

Scenario planning or visioning efforts attempt to achieve a regional consensus on desired future land use and transportation patterns, often focusing on longer timeframe (30 to 50 years) than the standard 20-year transportation planning horizon. MPOs and/or DOTs could be required to develop forecasts of GHGs under different transportation and land use scenarios, and to undertake a planning process with this broader focus. An example is California's Senate Bill (SB) 375, adopted in September 2008, which requires regional transportation plans to include sustainable communities strategies as part of the plan to achieve emission reduction targets. Federal policy would define clearly established linkages between the long-range vision and the existing long-range transportation plan.

This approach would encourage MPOs and State DOTs to be more visionary in their planning process, to look at a wider range of alternatives including land use patterns, and to include a broader range of nontraditional planning partners. Many metropolitan areas already are moving towards this type of approach. The willingness of many transportation agencies to adopt such an approach is likely to be low at least in the near-term, as most statutory authority for land use regulation lies at the local level and many agencies are not yet comfortable with the concept of being involved in local land use planning, even on a voluntary basis. Furthermore, strategies developed at a statewide or regional level are unlikely to be effectively implemented through local action if the region's various stakeholders are not willing participants. Local opposition known as "not in my back yard" often works against denser, more travel efficient development. Another barrier is fiscal zoning, where it is in the interest of municipalities to accommodate the most lucrative land uses, which have historically been considered low-density or auto-dependent. In this regard, showing examples of vibrant, mixed-used developments with substantial tax revenue can be of use. An alternative to a requirement for integrated planning would be to provide technical assistance and funding incentives for this type of planning.

GHG Reduction Targets

With appropriate congressional direction, the Federal government could either require State DOTs or MPOs to set their own GHG reduction targets (through the transportation planning process), or could set a national GHG reduction target (which could be uniform or apportioned to States and/or MPOs in different ways). State and regional transportation plans would be compared against these targets. The emissions targets would not identify the specific GHG reduction strategies to be implemented, but instead leave these to local planning agencies to determine. The question of how to enforce compliance with the targets would need to be addressed, including whether noncompliance would result in agencies being ineligible for certain highway funding incentives or larger impacts on broader highway funding.

State and regionally determined targets, if not mandatory, would likely be more acceptable to State and regional planners than national targets since States and regions would have the flexibility to set a target that they felt was achievable. On the other hand, they might lead to less aggressive targets being set than if the Federal government were to set targets nationwide. They also might lead to concerns about fairness if regions with more aggressive targets feel that they are shouldering a greater share of the GHG mitigation burden. Mandatory targets would be likely to encounter significant resistance from transportation agencies. A highly prescriptive process could lead to significant additional resource requirements to demonstrate future compliance with targets, and is not recommended by DOT.

Furthermore, care must be taken to recognize what is realistically achievable through planning actions. Transportation planners would not likely have influence over vehicle efficiency and fuel strategies and have limited control over even some nontechnology strategies such as land use planning.

Expanded Modal and Geographic Scope of Planning

Intermodal infrastructure planning can improve intermodal connections to make passenger and freight travel more seamless, allowing the utilization of the most efficient combination of modes for any particular trip. Planning and investment decisions can also shift travel to more efficient modes, higher occupancies, or higher freight tons/mile. Finally, planning can induce more efficient modal operations. One step in this direction is the recently passed Passenger Rail Investment and Improvement Act of 2008 (PRIIA), which established a requirement for a National Rail Plan which would consider GHG benefits as one of several criteria for prioritization. PRIIA also requires State rail plans, which are to be coordinated with the existing transportation planning process for highway and transit investments. The Federal Aviation Administration's NextGen program is another effort that could be useful in planning for other modes. NextGen seeks to develop environmental protection that allows sustained aviation growth. Finally, better incorporation of port operations in transportation planning processes could support more efficient ground operations, reduce truck traffic and emissions, and enable greater use of short-sea shipping. Efforts by the Maritime Administration (MARAD) and the Federal interagency group, Committee on the Marine Transportation System, are coordinating port infrastructure projects and leading national efforts to reduce congestion on the Nation's highways and rails by promoting the use of waterways and ports. Short-sea shipping already has been examined by regional interests such as the Port Authorities of New York and New Jersey and the Port Authority of Albany as well as by the International Mobility and Trade Corridor (IMTC) Project in the Pacific Northwest.

A new Federal planning program, implemented by DOT, could be created that is focused on multimodal, cost-effective, large-scale transportation strategies that improve mobility and reduce GHG from interstate travel. The goal of a national program would be to examine the synergies and tradeoffs among different interstate travel modes (including both passenger and freight transport), as well as identify policies and investments to leverage improvements to various modes that would reduce GHG. Such a program would focus on both passenger and freight transport, examining the National Highway System, intercity and freight rail, intercity bus service, air, ports, and intermodal terminals, and domestic waterways, and would coordinate with the existing statewide and metropolitan planning process.

Alternatively, a new planning structure and process could be established for megaregions or multistate geography. Megaregion or multistate planning efforts would encompass issues such as transportation and land use, low-carbon fuel strategies, multistate pricing policies, long distance freight movement, coordinated GHG reduction strategies, and coordination of large multi-jurisdictional projects. Planning at a megaregion or national-level may allow for some issues to be addressed more effectively than could be done for smaller geographic areas in isolation. However, the range of strategies that might be most effectively implemented at a megaregion level is limited (for example, major intercity transportation investments).

The development of planning structures at new scales is likely to encounter resistance from existing planning entities if they perceive an erosion of their authority, although multiregion planning has been embraced in some parts of the country (such as the East Coast's I-95 Corridor Coalition). Planning resource requirements would increase due to the establishment of new planning structures and the need for greater coordination among multiple entities.

Funding Incentives

Through the Federal-aid Highway Program and transit funding programs, the DOT provides funding to states and local governments to implement surface transportation programs and projects. Congress can change the structure of Federal transportation spending to prioritize GHG reductions, by directing funding towards specific types of planning activities or projects, or by establishing performance-based funding criteria to reward GHG emission reductions. Funding also can be targeted at GHG reduction activities in other modes of transport, including the rail, marine, and/or aviation systems. While future spending at any level could be better targeted toward strategies that prioritize carbon-efficient transportation projects, the Highway Trust Fund does not currently have enough resources to maintain current spending levels, let alone additional programs for low-carbon infrastructure. The DOT's *Conditions and Performance Report* finds that present highway and transit investment levels are insufficient to maintain the current conditions and performance of the system.[122] This, combined with deficits projected in the Highway Trust Fund, and competing goals for transportation spending, complicates the objective of aligning funding incentives with climate change goals.

Funding for GHG Planning Activities

Federal funding can be directed specifically towards planning for GHG reduction, including data collection, tool development, process refinement, strategy development, and analysis of GHG reduction strategies. The State or metropolitan planning agency would determine, through its planning process, which strategies should be implemented, considering the full range of benefits and impacts of each

---

[122] U.S. Department of Transportation. *Conditions & Performance: 2008 Status of the Nation's Highways, Bridges, and Transit.*

strategy. Funding could be targeted at those activities identified elsewhere in this report as having the greatest potential for GHG reductions.

This strategy would closely support technical assistance activities such as the application of improved planning tools and methods. It also would support any regulatory actions that require certain types of planning or analysis. It provides flexibility to State and metropolitan agencies in terms of what GHG reduction measures to actually implement, but does not guarantee that any specific measures will be undertaken.

Funding for GHG Reduction Strategies

Federal funding also could be directed at specific types of transportation projects that have been demonstrated to reduce GHG emissions. For example, the Congestion Mitigation and Air Quality Improvement Program (CMAQ) could be expanded or revised, or a new program created, to also provide funds to all states and MPOs to support projects that reduce transportation GHG. Funding also could be directed at other modes not normally funded through the statewide and metropolitan planning process. CMAQ already is used to fund a variety of nontraditional projects—such as intermodal rail freight projects and passenger ferries—and consideration of GHG impacts could be another selection criteria for such projects.

This type of approach would provide greater certainty that GHG reducing projects would be implemented, although the magnitude of such reductions would depend upon the types of projects funded and their level of use by consumers. While targeted funding can encourage specific activities, it can reduce flexibility to State and metropolitan agencies to meet locally defined goals and objectives. It also leaves fewer funds available to meet other needs, such as maintaining the condition and performance of the existing transportation system.

Performance-Based Funding

Performance-based approaches could range from a modest amount of funding to reward certain projects to completely changing how Federal funding is directed. The goal would be to reward activities resulting in the most cost-effective GHG reductions or areas achieving the greatest GHG reductions. A programmatic approach could, for example, take the form of a highway/transit formula factor that is based on transportation GHG per capita or is based on achieved reductions in transportation GHG per capita over time. In a performance-based funding approach, GHG is likely to be one of a number of performance measures, which may also include factors such as accessibility, safety, economic development, air quality, and livability. Resource decisions could be required to achieve certain performance targets in several areas.

A programmatic performance-based approach would allow each funding recipient the flexibility to choose the strategies that are most effective in widely varying circumstances. This approach could potentially be revenue-neutral overall for the Federal government, if there was enough revenue to achieve acceptable levels of performance in all categories, although it would cause shifts in funding among states and MPOs and transit operators. It may be difficult to achieve political support from individual states and regions that stand to lose funding if they cannot meet performance objectives. In addition, insuring sufficient funds are available to maintain the condition and performance of the transportation system in all regions may be a consideration. Furthermore, whether applied to individual projects or to overall programs, minimum requirements for technical analysis and oversight would be needed in order to verify that projected GHG reductions are actually likely to be achieved. This would add to the planning resource requirements of states and MPOs and the oversight responsibilities of DOT.

In each of the mechanisms discussed above, the Federal government should ensure that Federal agencies are working together to align funding policies and incentives and coordinate their programs. For instance, DOT, EPA, and HUD have established a Sustainable Communities Partnership. In addition, as transportation planning and funding decisions are made at all three levels of government—Federal, State, and local—Federal agencies must work in partnership with State and local governments, respecting the unique roles of each. Any Federal policies that are developed would need to keep in mind the needs of smaller MPOs, which often have fewer resources available for planning and therefore may find it more challenging to develop reliable, analysis-based performance measures for specific projects and programs.

4.2 PRICE CARBON

Mechanisms to price carbon emissions include:

- An increase in the **Federal motor fuels tax** to discourage carbon emissions.

- A **cap and trade** system, in which a limited number of GHG emissions allowances are traded in the market to cap overall emissions across all economic sectors;

- A **carbon tax** in proportion to the carbon content, or carbon dioxide-equivalent emissions, of the fuel.

There are three economic rationales for applying price signals:

- Carbon dioxide and other greenhouse gases can be considered an externality, and their full social costs (i.e., climate change-related damages) should be considered in business and consumer decisions; and

- Price signals harness market forces to identify and implement the most efficient and lowest costs mechanisms for reducing greenhouse gas emissions; and

- Economy-wide price signals help to solve the complex problem of incorporating life cycle emissions into a regulatory regime by including upstream emissions impacts arising from, for example, ethanol manufacture into the cost and price of the fuel.

In addition, revenue collected by pricing carbon can be invested in actions that further reduce carbon emissions.

Either a cap and trade system or a carbon tax approach would create a consistent set of prices across all sectors to encourage actions to reduce GHG emissions. Within the transportation sector, these actions would increase the cost of carbon-fueled transportation and would therefore create incentives for developing and purchasing more efficient vehicles and alternative fuels, as well as reducing travel and/or shifting to more efficient modes. An increase in the Federal motor fuels tax produces the same effects for transportation modes that use gasoline and diesel fuels. A motor fuels tax would not, of course, produce economy-wide emission reductions, nor would it necessarily produce relative prices for electricity and alternative fuels that accurately reflect their impact on emissions.

Increasing transportation fuel prices, from any source, has both immediate and longer-term effects. At sufficiently high levels, the immediate effect would be to reduce travel or freight shipments; to motivate more efficient operating practices; and to promote switching to less costly, and presumably more emissions-efficient, transportation modes.

The longer-term impact on fuel consumption and GHG emissions would be greater than the immediate impact, as transportation system users, fuel providers, and vehicle manufacturers can respond by manufacturing and purchasing more efficient vehicles, increasing the use of fuels with lower carbon content, and making more fundamental adjustments to activity patterns to reduce energy consumed in travel.

Further, the establishment at the Federal level of a cap and trade system, carbon tax, or increased motor fuel tax would create the expectation of long-term, sustained price increases—as compared to the unpredictable short-term price fluctuations seen in recent years. Achieving a policy environment with greater certainty in long term price trends will encourage long-run technological innovation and greater investment in more energy-efficient and reduced emissions vehicles and capital equipment in the transportation sector. A discussion of how travelers respond to both short and long-term price increases is included in Appendix A of this report. This section does not examine in detail the impact of carbon pricing on aviation or maritime industries.

Motor Fuel Tax

Increasing the Federal motor fuels tax would have a similar impact as a carbon tax (discussed below), but would only raise prices on motor fuels, rather than applying to all fuels based on carbon content. As with a cap and trade system or a carbon tax, a motor fuels tax would internalize the negative externality of environmental damage caused by burning fossil fuels and would provide a price signal to reduce fossil fuel consumption.

One study estimated that an increase in the Federal fuel tax of approximately 20 cents per gallon would reduce transportation GHGs by 2.3 percent in 2030 — with about three quarters of the effect coming from improvements in fuel efficiency and one-quarter from reduced VMT.[123] Levels equivalent to motor fuel costs in Western Europe, or an increase in the fuel tax of about $2.40 per gallon, could reduce transportation GHGs by 23 percent in 2030.

However, these higher prices would impose burdens on lower-income consumers and some businesses, and therefore would need to be implemented in concert with policies to address equity concerns. To give a sense of scale, fuel use per light duty vehicle averages 578 gallons per year.[124] This would decrease as vehicle efficiency improves.

Since an increase in the Federal motor fuels tax would only apply to the transportation sector, it could be adjusted to a higher level at which it would be likely to have a more substantial impact in the near term. Instituted primarily as a GHG reduction policy, this would entail imposing higher costs on the transportation sector than other sectors. However, unlike a cap and trade system or carbon tax, revenue raised through a motor fuel tax would have a strong precedent for being dedicated to transportation investments, as is currently the case with the existing 18.4 cent Federal motor fuels tax for gasoline and 24.4 cent tax for diesel.

An increase in the motor fuel tax could be scaled to generate sufficient revenue to cover shortfalls in the Highway Trust Fund and the gap between current transportation investment levels and the levels that the DOT *Conditions and Performance Report* estimates are necessary to maintain the conditions and performance of highways, bridges, and transit. Furthermore, it would maintain the "user-pay" principle of funding transportation investments from charging transportation users, rather than relying on general fund transfers to cover revenue shortfalls, which has been a recent trend with the general fund bailout of the trust fund in 2008 and 2009. Reinvesting the revenue in transportation infrastructure could yield mobility and congestion reduction benefits. Funding could be reinvested in strategies to further reduce GHG emissions from transportation, such as investments in alternative modes and system efficiency strategies. The motor fuels tax has the advantage of low administrative and compliance costs.

---

[123] Cambridge Systematics, 2009 (cited). The calculation was performed using a 1 cent per mile fee for the lower scenario, which equates to about 20 cents per gallon at the current average fuel efficiency of about 20 mpg for light-duty vehicles; and 12 cents per mile for the higher scenario, or about $2.40 per gallon. The calculations assume an elasticity of VMT with respect to operating costs of -0.45, as described in Appendix A.

[124] Federal Highway Administration, *Highway Statistics* 2006.

The treatment of alternative fuels under an expanded motor fuel tax regime is a key policy variable, particularly as renewable fuels penetration in the transportation sector has been mandated to increase substantially under the expansion of the renewable fuels standard by the Energy Independence and Security Act of 2007. Some alternative transportation fuels (such as ethanol and biodiesel) are eligible for a $0.45 per gallon excise tax credit. This credit is set to expire at the end of 2010. If alternative transportation fuels are exempted or continue to benefit from large tax credits, and there is no provision (such as exists currently in the renewable fuels standard) for limiting use of high life-cycle emissions alternative fuels, then some of the emissions benefits of the tax would be offset by increased life-cycle emissions from alternative fuels.

An increase in the Federal motor fuel tax is not proposed by the current Administration, given the economic recession. It has, however, been discussed by other policy actors in conjunction with the debate over the next authorization of surface transportation legislation. It is conceivable that the Federal motor fuel tax rate could be increased as a revenue generating mechanism to fill current and projected shortfalls in the Highway Trust Fund. Two recent Commissions, the National Surface Transportation Policy and Revenue Study Commission and the National Surface Transportation Infrastructure Financing Commission, recommended increased motor fuel taxes in the short term, by 10 cents per gallon, and with further adjustments for inflation. The Federal motor fuel tax rate was last increased in 1993 (prior to that time it was increased in 1990, 1986, 1983, 1961, and 1959).[125] Since that time, the motor fuels tax has lost much of its purchasing power due to inflation and other increases in materials and construction costs. A majority of State governments have continued to raise their motor fuel tax rates, with 32 of 50 increasing the rate since 1993; others have raised taxes from other sources (e.g., motor vehicle registration fees or excise taxes) to increase transportation revenue.

Cap and Trade

A cap and trade system provides, in principle, environmental certainty over the amount of emissions while at the same time using market forces to determine the most economically efficient actions to reduce emissions. The type of cap and trade system under consideration in Congress would require electric power generators, petroleum importers and refiners, and other large emitters of GHGs to hold allowances for each ton of their emissions. Allowances can be distributed initially through a government auction, free of charge, or a combination of the two. Entities that can reduce their emissions relatively cheaply will need to acquire and submit fewer allowances than entities that cannot reduce their

---

[125] Federal Highway Administration, *Highway History.*

http://www.fhwa.dot.gov/infrastructure/gastax.cfm.

emissions as cheaply. Entities can buy and sell allowances. In this way, the market system encourages the most cost-effective emissions reductions.

Entities that are upstream from the cap (e.g., coal mining companies), experience the cap and trade system as falling/rising prices for the products they sell. Entities that are downstream from the cap (in this case, airlines, railroads, and consumers) experience the cap in the form of higher energy and fuel prices.

Transportation GHG Reductions from Cap and Trade Limited in Near Term

In the transportation sector, the impact of a cap and trade system would be felt in the form of increases in fuel prices in the short term, rising over time as the cap tightens.[126] Fuel importers and refiners would be required to hold allowances for each ton of carbon dioxide equivalent that was contained in the fuel they sold. The EPA's modeling analysis of the cap and trade system proposed in the American Clean Energy and Security Act of 2009 estimated allowance prices of $13 per metric ton carbon dioxide equivalent in 2015, $16 in 2020, $27 in 2030, and $70 in 2050 in the core policy scenario. [127] The estimates are based on a system that would cover 85 percent of total U.S. GHG emissions, impose a cap starting in 2012 at 3 percent below 2005 covered emissions, and then

---

[126] Due to lack of available analysis there is little indication of the full impacts of a cap and trade system or tax on aviation or maritime industries.

[127] U.S. Environmental Protection Agency (2009). *EPA Analysis of American Clean Energy and Security Act of 2009, H.R. 2454 in the 111th Congress.* June 23, 2009, p3, p12, http://www.epa.gov/climatechange/economics/economicanalyses.html.

EPA models key uncertainties in different scenarios. Uncertainties covered in the scenarios include the degree to which new nuclear power is feasible, the availability of international offset projects, the amount of GHG emissions reductions achieved by the energy efficiency provisions in the bill, the impact of output based rebates to energy intensive and trade exposed industries. Across all scenarios modeled, the allowance price ranges from $13 to $24 per ton CO_2e in 2015 and from $16 to $30 per ton CO_2e in 2020, with the availability of international offsets having the largest impact.

This analysis falls within the range of previous EPA analyses as well as other entities' analyses on previous bills. Allowance prices are somewhat lower due to effects of Energy Independence and Security Act of 2007. For comparisons between modeling conducted by EPA, the Department of Energy, Massachusetts Institute of Technology, Clean Air Task Force, American Council for Capital Formation and the National Association of Manufacturers, see: Pew Center on Global Climate Change, "Innovative Policy Solutions to Global Climate Change: Insights from Modeling Analyses of the Lieberman-Warner Climate Security Act (S.2191)," May 2008.

The allowance prices shown reflect allowing domestic and international carbon offsets. Not allowing offsets would increase the allowance prices.

gradually reduce emissions to 17 percent below 2005 levels by 2020, and 83 percent below 2005 levels by 2050.

An allowance price of $15 per ton translates into an increase in the price of gasoline of $0.13 per gallon, based on its carbon content.[128] An allowance price of $30 implies an increase in gasoline prices of $0.26 per gallon. Price increases of other transportation fuels would rise similarly in proportion to their carbon content, as shown in Table 4.1.

---

[128] The price impacts shown are only due to the combustion of fuel. Since upstream and refining emissions will be included under cap and trade, the cost of these emissions will be reflected in the price seen by the consumer to some degree. The estimate also does not include general equilibrium effects, e.g., the dynamic effects on fuel prices from lowered demand for fuel as a result of the carbon price.

Table 4.1   Cap and Trade/Carbon Tax Price Impacts

	Gasoline	Diesel	Jet Fuel
Carbon Content kg CO₂/gallon	8.81	10.15	9.57
Allowance Price or Carbon Tax per ton CO2	**Gasoline $ per gallon**	**Diesel $ per gallon**	**Jet fuel $ per gallon**
$10	$0.09	$0.10	$0.10
$15	$0.13	$0.15	$0.14
$20	$0.18	$0.20	$0.19
$30	$0.26	$0.30	$0.29
$40	$0.35	$0.41	$0.38
$50	$0.44	$0.51	$0.48

Source: Carbon content from U.S. EPA (2007). *Inventory of Greenhouse Gas Emissions and Sinks: 1990 to 2005*, Annex 2.1.

$/gallon = kg CO₂/gallon * 1 metric ton/1000 kg * allowance price or carbon tax.

Note: The price impacts shown are only due to the combustion of fuel. Since upstream and refining emissions will be included under cap and trade, the cost of these emissions will be reflected in the price seen by the consumer to some degree. The estimate also does not include general equilibrium effects, e.g., the dynamic effects on fuel prices from lowered demand for fuel as a result of the carbon price. These price increases are low in comparison to the gas price increase of $2 per gallon experienced between 2004 and 2008,[129] which contributed to some leveling off of GHG emissions from the transportation sector but not steep declines.

According to a U.S. Department of Energy analysis of gas price increases over the last 10 years, gas demand was relatively inelastic, at -0.02 for prices over $2.50 per gallon.[130] This means a 100 percent increase in gas price is associated with a 2 percent reduction in gas consumption. Long run elasticities are greater,[131] as consumers have time to purchase more fuel efficient vehicles and move residences closer to work and other destinations. Researchers Small and

---

[129] U.S. Department of Energy, Energy Information Administration (2009). Monthly Energy Review, Table 9.4 Motor Gasoline Retail Prices, U.S. City Average, May 2009, http://tonto.eia.doe.gov/merquery/mer_data.asp?table=T09.04

[130] U.S. Department of Energy, Energy Information Administration (2008). "Short-Term Energy Outlook Supplement: Motor Gasoline Consumption 2008: A Historical Perspective and Short-Term Projections."

[131] Graham, D.J. and S. Glaister (2002). "The Demand for Automobile Fuel: A Survey of Elasticities." *Journal of Transport Economics and Policy* 36(1):1-26, January 2002.

Van Dender find short and long run elasticities of gasoline consumption with respect to price of -0.04 and -0.24, respectively.[132]

The modest increase in near-term fuel prices caused by a cap and trade system is not expected to spur large reductions in transportation GHG emissions. Under EPA's modeling conducted for draft cap and trade legislation, the electricity sector provides the vast majority of GHG reductions in the early years. Transportation and energy-intensive manufacturing see more emissions reductions in later years as the cap tightens and allowance prices rise.[133]

Analysis by the Energy Information Administration (EIA) of the cap and trade system in H.R. 2454, the American Clean Energy and Security Act (ACESA)[134] found reductions in transportation GHG emissions from ACESA of about 4 percent in 2030 relative to baseline emissions, or 85 million metric tons CO_2e.[135] This reduction results from a gasoline price increase of about 37 cents per gallon in 2030, corresponding to a $65 per tonne allowance price.[136] These estimates are for the basic case analyzed by EIA. Across the main cap and trade cases EIA analyzed, the transportation-related CO_2 emission reductions range from 2.6 to 8.5 percent (53 to 174 million metric tons).

This reduction results in part from a decrease in light duty and truck VMT of about 2 to 2.5 percent. Light duty vehicle fuel efficiency is only 0.3% higher in

---

[132] Small, K., and K. Van Dender (2007). *"Long Run Trends in Transport Demand, Fuel Price Elasticities and Implications of the Oil Outlook for Transport Policy,"* Discussion Paper No.2007-16. Small and Van Dender find that fuel price elasticities are a function of the share of gasoline expenditures in personal income, so that rising incomes tend to reduce price elasticities, while higher price levels tend to increase price elasticity. In the context of a cap and trade system, whether fuel prices become more or less elastic over time will depend on whether fuel prices (including allowance costs) rise more rapidly or slowly than per capita income. More information on price elasticities in this report is found in Appendix A.

[133] U.S. Environmental Protection Agency (2009). *EPA Analysis of American Clean Energy and Security Act of 2009, H.R. 2454 in the 111th Congress.* June 23, 2009.

http://www.epa.gov/climatechange/economics/economicanalyses.html.

[134] Passed the House of Representatives in June 2009. A companion Senate bill has not passed as of this writing.

[135] U.S. Department of Energy, Energy Information Administration (2009). *Energy Market and Economic Impacts of H.R. 2454, the American Clean Energy and Security Act of 2009.* http://www.eia.doe.gov/oiaf/servicerpt/hr2454/index.html. Figures cited here are for the basic case.

[136] The modeling performed by EIA finds that gasoline prices change under a policy scenario not only because of the direct impact of the allowance requirement, but also because of general equilibrium effects, such as a lower demand for fuels leading to slightly lower world crude oil prices.

the basic policy case than in the reference case. According to EIA, since all cases include the 35-mile-per-gallon CAFE standard enacted in the Energy Independence and Security Act of 2007, many of the most cost-effective vehicle efficiency options are adopted in all cases, including the Reference Case. EIA projects energy use by freight rail to decline 18 percent due to reduced volumes of coal shipments, because of the shift away from coal-fired power plants.[137] The EIA analysis does not analyze the additional ACESA provisions aimed at stimulating further advances in vehicle fuel efficiency and a more rapid penetration of vehicles that rely at least partially on electricity. If these additional provisions are successful, larger reductions in transportation sector emissions would be expected.

The EIA analysis did not include results for 2050, but these would be greater as allowance prices would increase over time, and responses to earlier price increases also are fully phased in.

The EIA analysis showed that between 80 and 88 percent of the energy-related CO_2 reductions in 2030 would come from the electricity generation sector of the economy, largely from reductions in coal usage. Figure 4.1 below shows estimated GHG emissions by sector.

---

[137] U.S. Department of Energy, Energy Information Administration (2009). *Energy Market and Economic Impacts of H.R. 2454, the American Clean Energy and Security Act of 2009.* http://www.eia.doe.gov/oiaf/servicerpt/hr2454/index.html.

Figure 4.1   Impacts of Cap and Trade on GHG Emissions by Sector

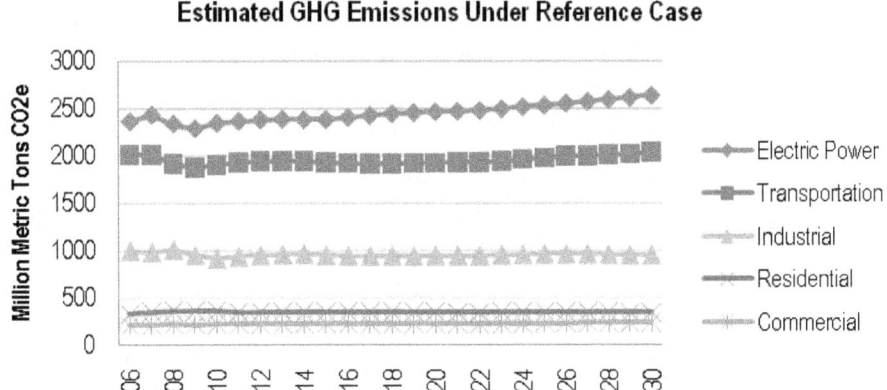

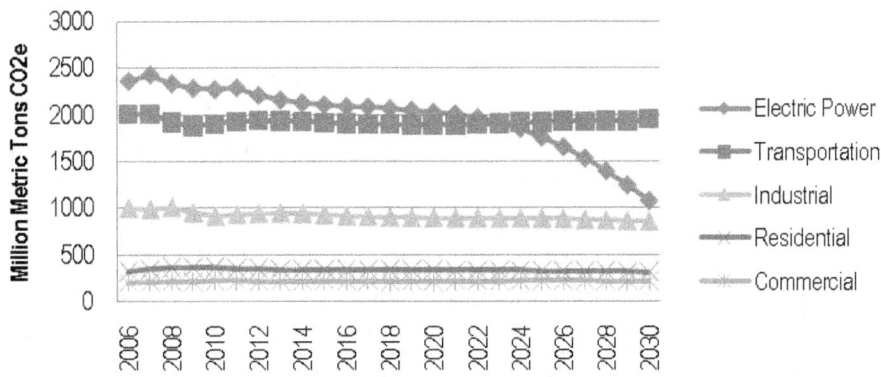

Source: U.S. Department of Energy, Energy Information Administration (2009). *Energy Market and Economic Impacts of H.R. 2454, the American Clean Energy and Security Act of 2009.* http://www.eia.doe.gov/oiaf/servicerpt/hr2454/index.html.

Note: The estimates for industrial, commercial, residential, and transportation do not include emissions from electricity as electricity emissions are counted under electric power.

In the long run, as rising allowance prices increase petroleum prices and cost effective solutions in other sectors have been exhausted, the impacts on the transportation sector are likely to increase. Depending on the availability of international offsets or lower-cost emission reductions in other sectors, it may be difficult if not impossible to achieve significant GHG reductions without GHG emission cuts in the transportation sector, since as Section 2.0 shows, the sector accounts for 29 percent[138] of U.S. GHG emissions.

---

[138] Including bunker fuels, or 28 percent not including bunker fuels.

Costs and Cost-Effectiveness

A cap and trade system encourages cost-effective emission reductions through the market mechanism. Those who can improve emissions for less than the price will do so, whereas those who cannot will purchase allowances. Collectively, emitters will therefore invest in those mitigation actions which are the most cost effective. The marginal cost of GHG abatement is equal to the allowance price.

Costs to the broader economy are estimated at less than one percent of U.S. gross domestic product (GDP) in 2030, according to EIA.[139] Average annual household consumption is estimated to decline in a range of $80 to $111 per household per year relative to the no policy case for the duration of the policy, according to EPA.[140] This represents 0.1 to 0.2 percent of household consumption. The costs include the effects of higher energy prices, price changes for other goods and services, impacts on wages, and returns to capital. Cost estimates also reflect the value of emissions allowances returned lump sum to households, which offsets much of the cap and trade program's effect on household consumption. A policy that failed to return revenues from the program to consumers would lead to larger losses in consumption.[141]

There are numerous other important factors that affect costs, such as the possibility of banking allowances, buying offsets from international carbon markets, "safety valves" to limit the price at which carbon is traded, minimum or "reserve" prices, and the reservation of some freely allocated allowances for new entrants. EPA estimates that if offsets were not allowed and emission reductions were achieved entirely through reductions in domestic covered emissions, permit prices would be 89 percent higher relative to the core policy scenario.[142] How the auction is designed and executed can also have a cost impact.

As for costs to the transportation sector specifically, an allowance price of $17 per ton CO_2e would increase motor gasoline costs for the U.S. transportation sector by $21 billion per year at current consumption levels.[143] However, carbon based fuel consumption levels would decrease with increased use of low carbon fuels,

---

[139] U.S. Department of Energy, Energy Information Administration (2009). *Energy Market and Economic Impacts of H.R. 2454, the American Clean Energy and Security Act of 2009.* http://www.eia.doe.gov/oiaf/servicerpt/hr2454/index.html.

[140] U.S. Environmental Protection Agency (2009). *EPA Analysis of American Clean Energy and Security Act of 2009, H.R. 2454 in the 111th Congress,* June 23, 2009, p.4, http://www.epa.gov/climatechange/economics/economicanalyses.html.

[141] U.S. Environmental Protection Agency, 2009 (cited).

[142] U.S. Environmental Protection Agency, 2009 (cited), p.3.

[143] Calculated from 2008 motor gasoline annual consumption level of 137.8 billion gallons reported by the Energy Information Administration.

efficient vehicles, improved system efficiency, and travel alternatives brought on by a cap and trade system or complementary policies.

Complementary Policies under Cap and Trade

A cap and trade system should, in theory, minimize the economic cost of a given level of emission reductions, and render additional strategies redundant. Under a cap and trade system, those sectors with the most cost effective emission reduction strategies available will reduce emissions and sell allowances to other sectors that are not able to reduce emissions as cost effectively. In theory, if it is more costly to reduce emissions from transportation than from electric power generators for instance, then it is economically efficient for most emission reductions to come initially from the other sectors while deferring significant transportation reductions until some time in the future. The environmental benefits of reducing directly emitted, long-lived GHGs such as carbon dioxide do not depend on where or how reductions occur.

However, if there are market failures that reduce the reaction to higher prices, then pursuing additional measures can lower implementation costs by compensating for market failures. There is evidence that some aspects of transportation, as well as other sectors, may exhibit market failures. For instance, consumers tend to undervalue fuel savings in vehicle purchase decisions.[144] That leads to the conclusion that a cap and trade system can serve as the central policy to guide cost-effective GHG reductions, while complementary policies (additional policies that work with the main cap and trade policy) also may be pursued where they can be demonstrated to lower implementation costs by compensating for market failures when they exist.[145]

A recent McKinsey report finds several strategies that could reduce GHGs while saving substantial sums of money.[146] These actions include insulating buildings and purchasing more efficient household appliances and vehicles that would save consumers more in reduced energy costs than it would cost them in up front costs. The National Highway Traffic Safety Administration (NHTSA) estimates that its recent model year 2011 fuel economy standard will reduce greenhouse gas emissions at a net savings of $1.5 billion over the time period

---

[144] Greene, D. L., J. German, and M. A. Delucchi (2009). "Fuel Economy: The Case for Market Failure." In *Reducing Climate Impacts in the Transportation Sector*, D. Sperling and J. S. Cannon, eds, Springer.

[145] There may be additional value in reducing emissions from a particular source or category of sources: for instance, reducing petroleum consumption may produce economic or national security benefits over and above the benefits of reducing the associated greenhouse gas emissions.

[146] McKinsey & Company (2007). *Reducing U.S. Greenhouse Gas Emissions: How Much at What Cost?*

2011 to 2030 as consumer fuel cost savings outweigh increased costs of vehicle technology.[147] If consumers undervalue energy efficiency, a carbon price signal alone will not elicit all cost effective emission reductions. Efficiency standards can compensate for market failures in consumers undervaluing energy savings. However, outside of the consumer sector, the evidence for market failure in commercial transportation operations such as trucking, aviation, maritime, or railroads is less compelling.

Models of cap and trade systems that include raising efficiency standards estimate lower allowance prices and consumer energy bills. These models also show that the availability of low-carbon technologies is critical to minimizing costs of GHG reductions.[148] As such, public investment in research and development can spur new energy technologies that reduce costs. This is particularly important to the extent that even with a price signal, private companies are hesitant to risk investing in uncertain, long term technologies or basic research.

The Government Accountability Office (GAO) surveyed a panel of 18 noted economists with expertise in climate change policies and found that 14 out of 18 believed that in addition to a cap and trade or carbon tax mechanism to establish a price for carbon, complementary policies such as investment in research or energy efficiency standards should be pursued.[149] However, the GAO report also emphasized the view of several panelists that energy efficiency standards can be economically inefficient and that vehicle fuel efficiency standards would be unnecessary in the presence of a robust mitigation policy to place a price on carbon. By not pursuing complementary policies for the transportation sector in addition to a cap and trade system, the sector may miss opportunities for investing earlier in low carbon technologies and mobility options, leading to higher costs in the future when allowance prices are high and these options are not available. As one expert put it, "Harnessing market forces is a very useful but probably insufficient strategy for mitigating transportation's GHG emissions. Even a carbon cap and trade system, as beneficial as it would be, would be hindered by the tendency of households to undervalue fuel economy. It would be unlikely to bring about an appropriate level of investment in long-term

---

[147] National Highway Traffic Safety Administration (2009). *Average Fuel Economy Standards, Passenger Cars and Light Trucks, Model Year 2011, Final Rule.* April 2009.

[148] Pew Center on Global Climate Change (2008). "Innovative Policy Solutions to Global Climate Change: Insights from Modeling Analyses of the Lieberman-Warner Climate Security Act (S.2191)"

[149] U.S. Government Accountability Office, 2008 (cited).

transportation energy technologies and would not guide important investments in transportation infrastructure and the built environment."[150]

However, complementary policies would need to be carefully chosen and designed to avoid overlap or increasing rather than decreasing implementation costs. In order to reduce total compliance costs, a complementary measure would need to be designed to compensate for a market failure that prevents higher allowance prices from inducing efficiency improvements. Also, the stringency or scale of the complementary policy would need to be set to generate net benefits at allowance prices equal to or lower than those generated by the cap and trade program.

Complementary measures could lower implementation costs but would not reduce national emissions levels

Because the national cap controls the total quantity of emissions, additional reductions in one sector caused by a complimentary policy permit increased emissions in another sector. The benefit of the complementary policy then is economic — lowering implementation costs — rather than environmental.

Implications for alternative fuels

A cap and trade program also would have distinctive effects on alternative fuels, with implications for complementary policies in this area. Synthetic fuels derived from coal or natural gas without carbon sequestration will become even more expensive than they are currently. Biofuels that depend on conventional fuel for process heat, and for petroleum-based transport for feedstocks or the fuel itself, will see some incremental cost increase.

In addition, biofuels may face increased competition for feedstocks under a cap and trade program. The grower of a 'fast rotation woody crop' suitable for conversion into cellulosic ethanol, for instance, could leave this crop standing and sell the sequestered carbon as an offset. Or, he could harvest the crop, and sell the crop into the electric power sector as biomass fuel to replace coal. Or, he could sell the crop to be processed into ethanol. Under cap and trade, market forces will move the crop to the 'best' use for achieving emission reductions.

Alternative uses for biomass as energy or sequestration crops under cap and trade may tend to raise the value of land and water, which will further affect biofuel economics.

Distribution of revenues could further affect transportation and consumers

Since the price elasticity of demand for fossil fuels is low, while the volume of fossil fuels is large, a cap and trade system that significantly constrains emissions could potentially raise large revenues. Allowances are valuable. If all, or a

---

[150] D. L. Greene and A. Schafer (2003). *Reducing Greenhouse Gas Emissions from U.S. Transportation.* Pew Center on Global Climate Change.

portion, of allowances are auctioned or sold at some approximation of their market value, then a constraining cap and trade system would actually generate large revenues.

The distribution of these revenues is a matter of intense political debate with groups arguing for revenues to be used for different purposes, including compensating consumers facing rising energy prices, funding energy research and development, investing in low carbon infrastructure, funding maintenance of existing infrastructure, helping energy intensive industries transition, and reducing fiscal deficits.

The transportation sector could be further affected if any revenues were reinvested in transportation to further reduce transportation emissions or compensate consumers facing higher fuel prices. For instance, Congress has considered investing cap and trade revenues in research on energy efficient vehicles, development of low carbon fuels, and investment in public transportation.

An analysis of equity is included in a recent Massachusetts Institute of Technology (MIT) study of alternative greenhouse gas control proposals.[151] The study found that carbon pricing impacts lower income groups more so than higher income groups. Assuming a carbon price of $15 per ton CO_2e, the increase in energy prices as a percentage of income ranged from 3.7 percent of income for the lowest income decile to only 0.8 percent of income for the highest income decile. The analysis also estimated the impacts of a "lump sum" rebate of all carbon revenues to all households, as the means to address equity issues. Rebating all revenues as a common lump sum would result in a 5.6 percent income gain for the lowest 10 percent of households to a 0.6 percent gain for the highest 10 percent of households. Thus, a full rebate in equal amounts to all households, of the proceeds of carbon pricing can eliminate the equity impacts on the lowest income groups. A similar effect would be expected at any level of price increase.

Research also has addressed other ways of offsetting the economic impacts of higher energy prices on low and moderate income households that would be associated with cap and trade or with a carbon tax. For example, income tax rebates or employment tax rebates have been suggested. One article proposes a payroll tax rebate for Social Security and Medicare taxes as an offset to carbon taxes.[152] The *Moving Cooler* study examined the equity implications of transportation pricing systems such as a VMT fee. The study found that pricing

---

[151] Metcalf, G.E., et al (2008). *Analysis of U.S. Greenhouse Gas Proposals.* MIT Joint Program on Science and Policy of Global Change, Report no. 160.

[152] Metcalf, G.E. (2007). "A Green Employment Tax Swap: Using a Carbon Tax TO Finance Payroll Tax Relief." Brookings Institution and World Resources Institute Policy Brief.

created inequities for lower income groups, but that these inequities could be addressed through reinvestment in highways, transit, system operations, commuter and ridesharing programs, and other transportation programs to improve mobility.[153]

One of the interesting side-effects of either a cap and trade system or a carbon tax is that rail freight traffic is expected to decline fairly substantially, due to the reduction in coal use as power plants shift to less carbon-intensive fuels.[154] While negatively affecting railroad revenue, this also should have the effect of reducing congestion and the need for investment on the rail system in corridors with large volumes of coal shipments, and potentially make it more feasible to shift some other freight from truck to rail. While reduced demand for coal would reduce rail traffic in some corridors, other corridors might see increases due to biofuels shipments by rail.

Carbon Tax

A carbon tax instituted at a comparable level to the permit price of a cap and trade system would have similar GHG reduction impacts, since a cap-and trade program looks like a carbon tax to energy consumers. The tax would need to rise over time and be adjusted to ensure the desired level of GHG reductions. Much of the discussion above regarding the impact of a cap and trade system on transportation also applies to the impact of a carbon tax on transportation. Under both, most reductions would come initially from the electricity generating sector, with more substantial reductions in transportation not occurring until the out years, when prices are higher. Market failures discussed above would similarly diminish the strength of the price signal.

Cap and trade and carbon tax policies have many similarities. Both strategies have been proposed as an alternative to a "command and control" approach in which the government would mandate how much individual entities could emit or what technologies they should use. Both are inherently market-based in that they send short and long term price signals that influence the decisions of consumers and businesses. Both correct a market failure, put a price on carbon, and take advantage of market efficiencies. Both policies impose a compliance obligation on a limited number of firms, can generate revenue, and may require special provisions to minimize impacts on low income consumers and industries dependent on fossil fuels. Both a carbon tax and a cap and trade system could be made more socially equitable by, in the case of a carbon tax, giving rebates to low

---

[153] Cambridge Systematics, Inc. (2009). *Moving Cooler: An Analysis of Transportation Strategies for Reducing Greenhouse Gas Emissions.* Urban Land Institute: Washington, D.C.

[154] U.S. Department of Energy, Energy Information Administration (2008). *Energy Market and Economic Impacts of S. 2191, the Lieberman-Warner Climate Security Act of 2007.* http://www.eia.doe.gov/oiaf/service_rpts.htm.

income households, and in the cap and trade system, compensating low income households using a portion of the revenue from the auction of allowances.[155] Finally, the market signal inherent to both can encourage greater change in behavior when alternatives to carbon intensive travel are available. When low carbon travel options are not available, the price poses a larger burden.

There are, however, key differences. A carbon tax offers more certainty regarding energy prices while a cap and trade system offers more certainty regarding overall GHG levels.[156] Many economists argue that a carbon tax would be more economically efficient because it would provide more economic certainty over prices than would a cap and trade system, as allowance prices would fluctuate. More certainty over prices provides industry with better information to guide investment decisions such as efficiency improvements and equipment upgrades. In addition, a carbon tax would allow more flexibility in emissions levels each year.[157] This can lower costs by, for instance, allowing more emissions in a year where a cold winter increased energy usage for heating. Total reductions are what is important rather than ensuring certain levels each year. A cap and trade system also can provide some flexibility in emission levels from year to year if it includes provisions that allow firms to borrow or bank emissions from year to year. Some contend that a carbon tax may provide implementation advantages such as greater transparency, a reduced administrative burden, and ease of modification.[158]

The effect of a carbon tax on alternative fuels is broadly similar to the impact of a cap and trade program, though the exact treatment of carbon sequestration under a carbon tax regime would influence the conditions under which biomass resources do or do not flow into biofuels.

A carbon tax has not been proposed by the current Administration or congressional leadership, although it has been advocated by some members of Congress.

---

[155] Pew Center on Global Climate Change (2009). "Climate Policy Memo: Cap and Trade v Taxes." March 2009.

[156] Pew Center on Global Climate Change, 2009 (cited).

[157] Congressional Budget Office (2008). "Policy Options for Reducing CO_2 Emissions."

[158] Congressional Research Service (2009). "Carbon Tax and Greenhouse Gas Control: Options and Considerations for Congress." February 23, 2009.

5.0 Policy Options

The results of this analysis demonstrate that the transportation sector has a significant opportunity to contribute to national reductions in GHG emissions.

GHG emissions from transportation can be reduced through four basic strategies: improving fuel efficiency; expanding the use of low-carbon fuels; improving the efficiency of the transportation system; and reducing the volume of travel that relies on carbon-based fuels.

Each strategy would require government policies in order to implement it and achieve GHG reductions beyond the business-as-usual scenario. This report does not provide recommendations, instead, it analyzes the potential of each strategy and the policy options for implementing the strategy.

In implementing these strategies, there are five broad categories of prospective policy action at the Federal level. They are:

1) **Efficiency standards** – options include fuel economy standards, low carbon fuel standards, and GHG emissions standards.

2) **Transportation planning and investment** – options include Federal technical assistance in integrating transportation and land use planning, and ensuring integration of climate change considerations into transportation planning and funding programs in order to prioritize GHG reducing strategies.

3) **Market-based incentives** – options include tax credits, feebates, subsidies, and vehicle miles traveled fees.

4) **Research and development** – options include research on advanced vehicle and fuel technology and research to develop data, tools, and decision-support to inform transportation planning and investment processes.

5) **Economy-wide price signal** – options include a cap and trade system or a carbon tax to establish a carbon price.

These approaches may be pursued jointly, and may have synergistic or reinforcing effects when implemented together (See section 3.9 - Key Interactions for more information). Implementing multiple actions can reinforce the level and pace of success in attaining steep GHG reduction goals. Integrated actions may be necessary to assure that, together, the steps taken to achieve GHG reductions also advance economic and societal goals. Approaches must also be evaluated carefully to ensure that there are no costly overlaps.

Table 5.1 shows which of the four strategies analyzed in the report these categories of prospective policy action support.

Table 5.1   Crosswalk between GHG Reduction Strategies and Categories of Policy Options

	Introduce low carbon fuels	Increase vehicle fuel efficiency	Improve transportation system efficiency	Reduce carbon intensive travel activity
Efficiency standards	x	x		
Transportation planning and funding programs			x	x
Market incentives	x	x	x	x
Research and development	x	x	x	x
Economy-wide carbon price	x	x	x	x

The policy options in the first category, economy-wide market pricing of carbon, would affect all sectors—electricity generation, industrial, commercial, and residential, as well as transportation. The policy options discussed in the other four categories are specific to the transportation sector, although the general categories apply to other sectors as well. For instance, the corollary to vehicle efficiency standards in transportation is household appliance efficiency standards in residential. An example of market incentives in the commercial sector would be tax credits for energy efficient windows. An example of the research and development category in electricity generation includes research into advanced solar technology. Government planning and funding programs also form a category of policy options in other sectors—for example, energy transmission line planning. The range of policy options specific to transportation within these broad categories are summarized below, after a discussion of general considerations.

Considerations

Some of the considerations that may be helpful in evaluating transportation policies in relation to GHG reductions include:

- Design GHG reduction policies that work in concert with other critical Federal priorities, including economic growth, mobility, and environmental sustainability, consistent with the DOT mission.

- Promote multimodal approaches to meet the Nation's growing transportation needs and environmental challenges.

- Promote more efficient use of existing transportation infrastructure through price signals that make travelers aware of externalities created by their transportation choices, including GHG emissions and others.

- Ensure focus on and progress in achieving GHG reduction goals through performance measurement, accountability, and transparency, consistent with the Administration's governing approach.

- Build on the lessons learned from State and regional initiatives, and promote innovation at the State, regional, and national levels.

- Pursue a coordinated portfolio approach to GHG reductions that taps the potential of multiple strategy groups. It is unlikely that any single strategy can achieve the steep GHG reductions being discussed.

- Collaborate with Federal partners to implement effective cross-disciplinary policies and programs since actions to reduce GHG emissions in the transportation sector will affect energy, environmental, and economic policies and programs.

- Promote options that are feasible in terms of technology, timely implementation, and cost effectiveness.

- Invest in research to develop breakthrough technologies and planning approaches to lower mitigation costs.

Efficiency Standards

Mechanism	• CAFE Rulemaking, National Program • Climate and energy legislation
Key Options	• Fuel economy standards • GHG emissions standards • Low carbon fuel standards
DOT Role	• NHTSA lead on establishing CAFE standards, in consultation with Federal partners • NHTSA and EPA joint rulemaking on National Program for harmonized vehicle fuel economy and GHG emission standards • Administration of fuel economy standards • Consultation with EPA/DOE on alternative fuels
Magnitude and Timing of Transportation GHG Reduction	• Modest to moderate in short-term, potentially very high in mid- to long-term

Since the 1970s, The DOT's National Highway Traffic Safety Administration (NHTSA) has promulgated fuel economy standards for light-duty vehicles. In early 2009, NHTSA set new fuel economy standards for the 2011 model year that will achieve an industry-wide combined fleet average fuel economy of 27.3 miles per gallon. That rule, for the first time, incorporated an analysis of GHG impacts associated with the new standards. Earlier this year, the Obama Administration directed NHTSA to conduct additional analysis of potential CAFE standards for future years, incorporating the most recent findings from ongoing analyses and studies. In addition, in May 2009, NHTSA and EPA issued a Notice of Intent stating their plans to work together closely to develop consistent, harmonized fuel economy and GHG emission standards for model years 2012 to 2016 under their respective statutory authorities. The NHTSA and EPA issued the proposed joint rule in September 2009 and the final rule in April 2010. Also, the National Academy of Sciences currently is conducting a study on fuel economy standards for work truck, medium, and heavy duty vehicles that NHTSA will rely upon to set future standards for these types of vehicles. Finally, in response to the new renewable fuel standard passed by Congress in December 2007, EPA conducted a life-cycle analysis of GHG emissions from biofuels. Each of these efforts contributes to the research literature and informs the development of this policy option.

Vehicle Standards

As reported in the technical findings of this report, more efficient vehicle technology shows strong potential to reduce GHG emissions across all transportation modes, assuming progress in ongoing research, development, and deployment. Strong standards for fuel efficiency can help achieve GHG reductions in the near- to mid-term, as the vehicle fleet turns over, by decreasing the amount of carbon consumed per mile of travel. Equally important, these standards would help stimulate the research and development that will be required for future progress. Because light-duty vehicles account for 60 percent of U.S. transportation GHG emissions and evidence shows consumers do not fully incorporate fuel savings into purchase decisions, fuel economy regulations for light duty vehicles can have a large impact on U.S. emission levels.

The Energy Independence and Security Act of 2007 (EISA) requires a fleet average light-duty vehicle fuel economy of at least 35 miles per gallon by 2020 for light-duty vehicles.[159] The NHTSA issued a one-year rule that sets fuel economy standards for Model Year (MY) 2011 at an industry-wide, combined fleet average of 27.3 miles per gallon. In April 2010, NHTSA and EPA issued a joint rule to establish CAFE standards and vehicle GHG emissions for passenger cars, light-duty trucks, and medium-duty passenger vehicles built in model years 2012 through 2016. The intention is to have a coordinated program, or National Program, that can achieve substantial improvements in fuel economy and reductions of GHG emissions from the light-duty vehicle part of the

---

[159] The EISA mandated that the model year 2011-2020 CAFE standards be set sufficiently high to ensure that the industry-wide average of all new passenger cars and light trucks, combined, is not less than 35 miles per gallon by MY 2020.

transportation sector, based on technology that will be commercially available and which can be incorporated at a reasonable cost. In addition to NHTSA's fuel economy standards under the Energy Policy and Conservation Act (EPCA), EPA proposed the first ever Federal emissions standards for GHGs using its authority under the Clean Air Act (CAA). The intent of the National Program is to allow auto manufacturers to build a single light-duty national fleet which satisfies requirements under both programs and which provides significant reductions in both light-duty vehicle oil consumption and GHG emissions.[160] Taken together, the NHTSA and EPA standards are expected to result in an industry-wide, combined fleet average of an estimated 35.5 miles per gallon by MY 2016. Consideration can be given to further increasing standards through subsequent legislation and regulation addressing longer-term efficiency targets beyond 2016 and 2020.

In addition to regulating light-duty vehicles, Congress has considered and required DOT to study and issue standards for work truck, medium, and heavy-duty vehicles whose emissions have grown at three times the rate of light duty vehicles since 1990, and now account for 20 percent of U.S. transportation GHGs.[161] The NHTSA will take action based on findings of a study in progress by the National Academy of Sciences and subsequent study by DOT.

It should be noted that NHTSA is pursuing other measures that support fuel efficiency. In response to EISA, NHTSA published a March 2010 final rule on a rating and labeling system for replacement tires that rates tire rolling resistance and efficiency—another strategy designed to improve fuel efficiency of vehicles. Other initiatives underway by NHTSA include a vehicle rating program for consumers (final rule scheduled for December 2011) and development of a consumer education program on fuel savings and alternative fuel vehicles, scheduled to roll out in December 2011.

Finally, standards may also be considered for rail, air, and marine modes, although such standards would likely be more difficult to apply. An alternative to fuel economy standards would be to require the use of certain technologies that have been proven feasible and cost-effective in reducing fuel consumption (such as drag reduction on trucks or trains).

Fuel Standards

A low-carbon fuel standard focuses on carbon levels rather than on fuel economy, and allows fuel suppliers to determine how to cost-effectively meet the carbon standard through combinations of fuel strategies. This technology-neutral approach rewards the lowest carbon results without choosing winners

---

[160] http://www.whitehouse.gov/the_press_office/Remarks-by-the-President-on-national-fuel-efficiency-standards/.

[161] See Vol. 1, Sec. 2.0.

and losers among technologies used in the reduction of carbon intensity in transportation. By providing certainty over future demand for low-carbon fuels, it also encourages vehicle manufacturers to design vehicles that support the use of such fuels (such as bio-fuel or flex-fuel vehicles capable of running on both gasoline and ethanol).

There are several issues to consider when contemplating a low-carbon fuel standard:

- **Overlap.** A low-carbon fuel standard would overlap with the existing volumetric renewable fuel standard, which mandates that 36 billion gallons of renewable fuel be in use in the U.S. by 2022. This could lead to increased compliance costs incurred by regulatory agencies and producers.

- **Cost.** Fossil fuel producers must acquire a certain amount of a low-carbon fuel in order to earn the right to sell a high-carbon fuel (i.e., gasoline). The extra cost (if any) of the low-carbon fuel is thus added to the cost of the high-carbon fuel. If low-carbon fuel supplies are insufficient, prices of both low-carbon and high-carbon fuels would rise until: 1) additional low-carbon fuel supplies are forthcoming; or 2) consumption of high-carbon fuels is suppressed.

- **Qualifying low-carbon fuels.** Establishing a low-carbon fuel standard would require the Federal government to quantify the life-cycle greenhouse gas emissions from different types of fuels. Particularly challenging is estimating indirect emissions from land-use change induced by producing additional quantities of crops for biofuels. This also is a challenge for the existing renewable fuel standard, as it requires fuels to have emissions a certain percentage below that of conventional fuels in order to qualify.

- **Coverage.** Congress must decide if only on-road gasoline and diesel are covered or also jet, marine, or other fuels. The broader the scope of the rule, the more opportunity for substituting low-carbon fuels and the lower the cost per gallon in cross-subsidy for a given volume of low-carbon fuels. Congress also must decide if electricity qualifies as a low-carbon fuel. Even electricity generated from fossil fuels might qualify as a low-carbon fuel if the efficiency of generation, transmissions, storage, and conversion to power to move the electric vehicle produced emissions lower than the life-cycle emissions from fuel production and combustion in a gasoline-powered internal combustion engine.

- **International effects.** Imposing a low-carbon fuel standard in the U.S. alone could lead to companies "shuffling" production and sales in ways that do not reduce emissions.[162] For example, low-carbon fuels could be sent to the U.S. with higher-carbon fuels going to other countries without such requirements.

Transportation Planning and Investment

Mechanism	• Surface Transportation Authorization bill
Key Options	• Federal technical assistance in integrating transportation and land use planning • Range of options for ensuring integration of climate change considerations into transportation planning and funding programs • Speed limit reductions • Operational improvements and pricing • Investments in public transportation and pedestrian facilities
DOT Role	• Development of Authorization legislation proposals • Technical support to Congress
Magnitude and Timing of Transportation GHG Reduction	• Modest in the short-term, moderate in the mid-term and ongoing

GHG Considerations in Planning and Investment Processes

The Federal government invests billions of dollars annually in support of transportation infrastructure. How these massive investments are directed in the coming decade will shape the future of the Nation's transportation system and will have profound and lasting impacts on the sector's level of GHG emissions.

As discussed in Section 4, the planning process inherently must incorporate a wide range of considerations—including mobility and accessibility objectives, safety, economic development, resource constraints, environmental sustainability, and land use—to shape a community's ideal long-range vision and identify the priority projects, programs, and strategies for achieving that vision. The GHG impact of transportation programs is a key element that can be addressed in this process, and indeed is more effectively addressed as part of a system-level planning process than at an individual project level.

The DOT is proactively engaged with States and MPOs to develop best practices in incorporating climate change into their planning processes. This support to

---

[162] Sperling, D. and S. Yeh. "Transforming the Oil Industry into the Energy Industry." *Access*, Spring 2009.

State and regional agencies includes provision of guidance and technical support, review and input of State climate action plans, and sponsorship of peer exchanges to assist agencies in developing effective planning practices under existing regulations. The DOT will continue to provide leadership to promote the incorporation of climate change considerations in planning and investment decisions.

To ensure that investments are aligned with GHG reduction objectives as well as other transportation goals, funding criteria and performance measures could be set that promote and fund passenger infrastructure and service expansion based on mode-neutral GHG reduction performance measures. Such investments might include congestion relief, urban transit, intercity bus and rail, intermodal passenger facilities, and nonmotorized facilities. Similarly, funding criteria and performance measures could be applied to freight infrastructure and service expansion, including investment in rail, intermodal facilities, and port access and operational improvements that would achieve GHG reductions.

The ability for States and regions to achieve effective and sustainable GHG reductions is grounded in the long-range planning and investment process. Within Federal guidelines, States and MPOs develop long-range plans (at least 20 years) that address mobility and environmental objectives, and prioritize near-term investments within financial constraints. The scope of considerations addressed in the transportation planning process has continuously expanded as communities and transportation professionals recognize the important interactions among transportation decisions, land use planning, environmental sustainability, conservation planning, public health, and economic development. The current surface transportation authorization bill, SAFETEA-LU, requires State DOTs and MPOs to consult with resource agencies and make use of available conservation maps and ecological data as transportation plans are developed. Federal transportation and resource agencies are working collaboratively to promote integrated planning processes that support effective long-range planning processes that meet multiple environmental and transportation objectives. *Eco-Logical*, which was issued in 2007 as a joint policy report of the DOT and several Federal resource agencies, outlined principles to achieve an improved integrated planning approach. The current partnership of DOT, HUD, and EPA is addressing housing and development, environmental, and transportation objectives through integrated planning.

A range of strategies to ensure climate change considerations are integrated into transportation planning processes are under debate, as discussed in this report's section on the planning process. Options range from including climate change as a planning factor, to requiring States and MPOs to develop strategies for reducing transportation GHGs, to establishing mandatory GHG reduction targets. Each option will have differing levels of impact on GHG emissions and on level of effort required. Federal transportation funding programs can provide incentives for GHG reduction. Another option is to align Federal funding for transportation infrastructure with performance-based criteria, including climate change objectives.

States and regions have varying capacity to address climate change in their planning process, and the availability of appropriate data and model outputs is insufficient to support robust analysis. Therefore, Federal action would be helpful to support for the research, technical support, and capacity development that transportation and planning agencies need to conduct planning. Federal programs can be created to develop and provide improved data, tools, and technical assistance for GHG planning and analysis. Demonstration projects at megaregion, regional, State, or local levels can build knowledge and capacity of effective GHG reduction strategies.

Operations Investments to Improve System Efficiency

If properly directed, investments in infrastructure and system operations can improve the efficiency of the transportation network, thereby improving the quality of passenger travel and goods movement. This improved efficiency also would help reduce congestion and travel delay, resulting in some GHG reduction for those segments of the system. System efficiency investments can achieve GHG benefits through reducing speeds to optimal energy-efficiency levels and enhancing efficiency through operational improvements.

Operational Improvements

GHG emissions can be reduced by improving the efficiency of the transportation system. System efficiency strategies also have substantial cobenefits that support other DOT goals, particularly in congestion management, air quality, and streamlining goods movement. These strategies vary significantly in cost and ease of implementation. Examples identified in this report include:

- Funding of integrated corridor management/advanced traffic management to keep traffic flowing at optimal speeds; and

- Support for research, evaluation, deployment and infrastructure development for advanced vehicle and information technology (e.g., advanced traveler information systems, vehicle-infrastructure integration programs).

System efficiency strategies can be evaluated based on their efficacy in addressing GHG emissions, along with other criteria that are currently applied (such as mobility, safety, and air quality). Ensuring integration of climate change considerations in transportation planning and funding would enable increased support to be directed to climate-friendly system improvements and technology applications. In order to do so, further research and guidance on GHG analysis methods would be required, particularly on the issue of induced travel (i.e., additional travel taken in response to improved travel conditions).

Two policy approaches under discussion are the expansion of the current CMAQ program or creation of a new program directed specifically to projects that meet

GHG reduction performance criteria. These options—which could address several strategies (e.g., system efficiency, travel behavior, climate-focused planning)—merit consideration during development of legislative proposals for authorization.

The key legislation mechanism for this policy option is the surface transportation authorization bill.

Speed Limit Reductions

Setting lower speed limits could have an immediate and significant impact on GHG reductions as well as yield substantial safety and air quality co-benefits. National reduction in highway travel speeds could stimulate research in alternatives for cost-effective high-speed freight and passenger transport for long-distance travel.

However, there are both political and practical hurdles to implementing this strategy. Public resistance is likely to be high, and an aggressive education program and strong political leadership would be required to gain broad support. Delay costs could be incurred in goods movement and passenger travel. Motor carriers also may be reluctant to accept lower speed limits if the speed limits significantly reduce driver productivity, although they would also benefit from reduced fuel costs (as evidenced by the fact that many long-haul trucks already are governed to keep highway speed down and improve fuel efficiency). In addition, this strategy would require enhanced enforcement and could impose considerable costs on States to pay for increased traffic monitoring and enforcement.

Programs to Reduce the Demand for Carbon-Intensive Travel

These policies reduce the demand for carbon-intensive travel by facilitating the use of alternative modes such as public transportation, carpooling, walking, biking, and by reducing the need for long trips by integrating land use and transportation planning. These strategies—implemented through a combination of pricing, funding, and policy incentives—can have modest near-term impacts in reducing GHGs. Over the longer-term (to 2050 and beyond), shifts in land use that promote more compact development, combined with significant investments in transit and rail capacity, have the potential to achieve significant and cost-effective GHG reductions. Examples of strategies to reduce travel activity are:

- Federal regulations, technical assistance, and funding to support integrated transportation and land use planning;

- Investment in low carbon modes such as public transportation, walking, and biking;

- Federal guidance and support to local and regional governments in promoting worksite trip reduction through tax credits and incentives to employers and workers, as well as trip reduction programs for community-wide travel to nonwork destinations (e.g., universities, schools, special events). Example strategies include telework, compressed work week, transit benefits, and rideshare support;

- Federal guidance for highway design and reconstruction that require accommodation of pedestrians and bicycles and application of context-sensitive design principles in all projects;

- Support for pricing initiatives (e.g., congestion pricing, State VMT fees, parking fees); and

- Promotion of eco-driving, which aims to change individuals' driving behavior and techniques to reduce fuel usage and GHG emissions without changing the number or length of trips.

The actions and strategies under this category have modest to moderate impacts on GHG reductions, but transit, carpooling, bicycle, pedestrian, and other ridesharing programs also serve to enhance access to jobs and opportunities, particularly for low-income groups. Expanding traditional programs in this category and implementing new initiatives can address equity issues that would be created by actions that increase the costs of travel. Policy actions that increase costs — such as economy-wide or other pricing measures — must be offset with policy actions that provide desirable travel choices for low-income and auto-dependent users.

Legislation also could mandate that States authorize or require companies to offer Pay-as-You-Drive (PAYD) insurance. This would make insurance costs more transparent and provide an incentive for drivers to limit unnecessary travel. Pay-as-You-Drive would have a modest to moderate additional impact on GHG reductions in the mid-term.

Market-based Incentives

Mechanism	• Surface Transportation Authorization • Climate or energy legislation
Key Options	• Tax credits, feebates, subsidies, vehicle miles traveled fees
DOT Role	• Development of Authorization legislation proposals • Technical support to Congress
Magnitude and Timing of Transportation GHG Reduction	• Moderate, depending on pricing level, in the mid-term; potentially very strong in long-term

The use of sector-specific market signals—including both incentives (such as rebates and tax deductions) and pricing mechanisms (such as taxes and fees)—can encourage consumers to more quickly adopt less carbon-intensive vehicles, technologies, and travel behaviors. By altering consumer demand, these market signals can spur more rapid private sector research and development of advanced technologies and decrease demand for more carbon intensive approaches to mobility.

Fuel Incentives

At the national level, the Energy Independence and Security Act of 2007 (EISA) requires a certain volume of renewable fuels (e.g., cellulosic ethanol, biomass-based diesel), with increasing volumes from 2006 through 2022. Along with these requirements, the law provides credits for producing additional renewable fuel. EISA sets production benchmarks of 36 billion gallons of renewable fuel by 2022, of which 21 billion gallons must be advanced biofuel. The bill (in Section 712) authorizes the use of grants and government-backed loans to assist manufacturers in converting plants to encourage domestic production and sales of efficient hybrid and advanced diesel vehicles, as well as components of those vehicles. The bill also extends the Flexible Fuel Vehicle Credit Program (Section 109), allowing manufacturers to take fuel economy credit for dual-fueled (i.e., flex-fuel) vehicles for their corporate average fuel economy (the credit phases out in 2019).

National targets and incentives for manufacturers such as these can be supported by market signals at the consumer level that would encourage purchase of high-efficiency and non-carbon-based vehicles, as well as retrofits of existing vehicles with technologies that improve fuel efficiency.

Rebates, Fees, Tax Incentives

A series of incentives and disincentives can be implemented to promote the rapid market penetration of low-GHG emission vehicles. These price signals would encourage consumers to purchase new technologies, thereby creating a stronger market demand that spurs production. The success of tax incentives for hybrid vehicles provides a good example of this approach. Tax credits and feebates can target individual households to encourage the purchase of low-carbon vehicles. Feebates impose a fee on purchasers of inefficient vehicles and provide a rebate to purchasers of high efficiency vehicles. Similarly, tax incentives can be implemented to encourage businesses to invest in new technology vehicles.

Fee structures that provide disincentives for the purchase of high-carbon vehicles—such as graduated registration fees that increase based on carbon-intensity and fees on the purchase of high-carbon vehicles—also provide pricing signals that reward GHG-reducing consumer behavior.

To be most effective, the design of these consumer incentives/disincentives should be technology-neutral, based on GHG emission-level metrics that are not prescriptive regarding the specific type of fuel or vehicle employed to achieve those targets.

Implementation of other technologies—such as anti-idling and drag reduction technologies for freight vehicles—could be similarly encouraged through rebates, retrofit programs, and tax incentives. As an example, the Energy Improvement and Extension Act (EIEA) excludes certain idling reduction devices and advanced insulation from the Federal excise tax.[163] Programs such as SmartWay℠, aimed at encouraging the freight industry and truckers to adopt efficient technologies, have been effective and can be continued. Retrofits in the heavy-duty and other sectors can achieve some immediate efficiency gains, helping to overcome the problem of slow fleet turnover in these sectors that lengthens the time needed to realize the full benefits of new energy-efficient vehicles.

Several legislative mechanisms can be used to implement these incentives and to encourage existing successful programs. These include language in the surface transportation bill and in climate and energy legislation.

Motor Fuel Tax Increase/VMT Fee

Sector-specific pricing measures also can play a significant role in reducing the carbon-intensity and volume/frequency of travel. In particular, increased motor fuel taxes or VMT taxes would provide incentives to travelers to reduce trip lengths/frequencies and shift to less carbon-intensive modes. However, significant price increases would be required to make a significant impact on travel, and such increases are likely to be met with considerable public opposition. Strong Federal leadership and bipartisan support would be required to achieve a meaningful fuel tax increase.

Increased fees on travel also create important equity concerns that must be addressed. As the cost to consumers of travel is raised, the burden would be placed disproportionately on lower-income groups, and on some residents of rural areas. A mechanism for compensating transfers or equivalent investments to expand travel options that improve access to for these groups may be required.

---

[163] Energy Improvement and Extension Act of 2008 (PL 110-343), Section 206.

Research and Development

Mechanism	• Climate or energy legislation • Surface Transportation Authorization bill
Key Options	• Research on advanced vehicle and fuel technology • Development of data, tools, and decision-support to inform transportation planning and investment processes • Policy-oriented research to analyze interactions between GHG reduction policies, costs, benefits, implementation considerations, and equity aspects
DOT Role	• Development of Authorization legislation proposals • Technical support to Congress • Partnerships with U.S. Climate Change Technology Program and U.S. Global Change Research Program • Research partnership with DOE and EPA
Magnitude of Transportation GHG Reduction	• High in the long-term

A strong Federal program of interdisciplinary research and technology deployment can advance the effectiveness of the transportation sector in addressing climate change. These areas are:

- Basic and applied research in non-carbon-based fuels and advanced high-efficiency vehicles across all modes;

- Advanced research to identify "break-through" technologies that could alter the course of vehicle technology;

- Research to develop data, tools and decision-support to inform transportation planning and investment processes in developing GHG reduction strategies, including tools to support regional and local modeling, long-range scenario planning and program development;

- Climate research that advances understanding of the relationships between transportation GHG emissions, climate changes, and climate risk analysis and adaptation strategies at the regional level;

- Research and deployment of information technologies to support system efficiency and maximize operational and travel behavior strategies;

- Policy-oriented research to conduct detailed analysis on the interactions between GHG reducing transportation policies, the costs of policy measures, and implementation considerations; and

- Research on the equity implications of transportation GHG reduction policies and technologies, and on ways to mitigate or avoid any negative equity impacts.

Technology research can be supported through an ongoing program that funds technology-neutral advanced research and development programs for all modes. The program should be designed to address both near-term improvements—those that can be reasonably achieved and implemented within the next 10-to-15 years—and technologies that can be expected to achieve break-through advancements in the longer run. Support for technology research should focus on outcomes rather than on selecting winners and losers. As noted in a recent MIT report, Federal policies should be designed to "Push development and deployment of appropriate technologies—and generate market pull for those technologies—through policies that reinforce each other through synergies. Incentives should be for outcomes, and not be focused on particular technologies that put other vehicles with low fuel use and emissions at a competitive disadvantage."[164]

In addition, a specific focus on technology deployment is required. This would include identification of barriers to implementation that may be addressed through Federal capital investments in infrastructure, technical support and funding incentives for State and regional agencies, and public-private investment strategies.

Past Federal support of public-private partnerships for vehicle and fuels research has been very effective. The Partnership for a New Generation of Vehicles, though later suspended, was successful in producing diesel hybrid prototypes; however, these innovations never reached the marketplace. A commitment to serious and sustained investment in research and development, combined with market pulls such as carbon pricing, would support industry in long-term development and deployment of promising technologies. Further, these investments support creation of a retooled manufacturing sector that can meet the needs of the future while being a major driver in the Nation's economic recovery.

The Energy Independence and Security Act of 2007 establishes seven bioenergy research centers and provides grants for the research and production of advanced biofuels. Overseen by the Department of Energy, research grants totaling $500 million are authorized for FY 2008 to 2015. In addition, funds are authorized for competitive, university-based, research awards.

Several appropriate legislative mechanisms can be considered to achieve this research agenda. Support of joint research programs that engage both Federal

---

[164] Bandivadekar et al. (2008). *On the Road in 2035: Reducing Transportation's Petroleum Consumption and GHG Emissions.* Laboratory for Energy and the Environment, Report No. LFEE 2008-05 RP, Massachusetts Institute of Technology.

and industry research programs—such as DOE's FreedomCAR and Fuel Partnership—can be continued. Upcoming climate and energy legislation could include funding of a substantial joint research program to accelerate advanced research efforts. This basic research would set the stage for private investments in technologies with strong market potential.

The need for extensive climate research—including a dual focus on both mitigation and adaptation—is an ongoing national priority. For transportation as well as other sectors, a strong component of this research should be focused on the information needs of national, State, and local planners and system managers. Federal interagency research partnerships—most notably the U.S. Climate Change Technology Program and U.S. Global Change Research Program—include active participation and direction from DOT and other program agencies. This close collaboration helps ensure that Federal research is conducted that addresses the specific data and technical needs of transportation decision-makers.

Economy-Wide Price Signal

Mechanism	• Climate or energy legislation
Key Options	• Cap and trade
	• Carbon tax
DOT Role	• Technical support to Congress
Magnitude and Timing of Transportation GHG Reduction	• Modest in near-term, moderate in mid-term, potentially strong in long-term

An economy-wide price signal could be established through a cap and trade system or a carbon tax. Cap and trade legislation is the primary policy option currently under discussion. Information on carbon taxes is found in Section 4. Legislation to establish a cap and trade program would support all strategies by creating a price signal for carbon that incorporates the negative externalities of carbon-based fuel use. An increased price for carbon would provide incentives for consumers and businesses to minimize carbon-based fuel consumption. This would help drive the development of cost-effective responses (e.g., technology development, travel behavior changes) that would reduce GHG emissions. The implementation of carbon pricing—assuming a sufficiently strong price is established—would result in reductions in fuel consumption and an ongoing shift to non-carbon-based fuels and technologies across all sectors.

In the near-term, however, the direct impact of an economy-wide cap and trade pricing scheme on the transportation sector is expected to be limited compared with the impact on other sectors, especially the electricity generation sector. By definition, pricing approaches for carbon stimulate reductions in those sectors for which the most cost-effective alternatives can be identified. If the cost of carbon fuels to transportation is insufficiently high, significant transportation responses would be delayed until prices further rise. Further, market deficiencies can be expected because drivers tend to undervalue fuel prices in their purchase decisions.[165] Whereas cap and trade prices would likely be low initially and increase over time, a high cost impact is needed to make research, development, and adoption of new technology worthwhile to the private sector.

While some argue that a cap and trade system should be allowed to function on its own to encourage the most cost-effective strategies economy-wide, others hold that complementary transportation strategies should be concurrently pursued. In the face of a potential delay in transportation sector response to economy-wide signals, a multi-policy approach has several advantages. Implementation of strategies targeted directly to the transportation sector can stimulate technology development that would support transportation efficiencies over time. This early investment would be more cost-effective than deferred technology research in coming decades, and would better position U.S. producers to compete in the global economy. Further, complementary strategies can help counter the market failures anticipated by consumers' response to increased fuel costs.

Congress is considering proposals to achieve economy-wide carbon pricing. Of particular importance to the transportation sector in the design of a cap and trade program is the use of auction revenues. A portion of auction revenues or of the revenues from a broad carbon tax could be allocated to transportation in two areas: 1) multimodal infrastructure and system development that supports additional GHG reductions and provides benefits to all groups to remedy the equity issues created by higher energy prices; and 2) research to develop advanced transportation technologies and fuels. In addition, recognizing that any economy-wide market-based measures would have impacts on low-income populations, any carbon-pricing program could include provisions to address equity disparities through individual tax credits, carbon-efficient travel subsidies, lump-sum rebates to low-income households, and public transit initiatives.

---

[165] Greene, D. L., J. German and M. A. Delucchi (2009). "Fuel Economy; The Case for Market Failure." In *Reducing Climate Impacts in the Transportation Sector*, D. Sperling and J. S. Cannon, eds, Springer.

6.0 Conclusion

The ingenuity of transportation planners and engineers has produced a vast network of transportation infrastructure and services to support the mobility and economic vitally of the Nation. However, our historic approach to transportation and land use development has created an energy-intensive system dependent on carbon-based fuels and individual vehicles.

Our national talents and resources must now focus on shaping a transportation system that that serves the Nation's goals, including meeting the climate change challenge. This will require aligning funding programs and incentives so that national investments are targeted to achieve GHG reductions, while continuing to meet mobility and accessibility objectives for both passenger and freight travel across all modes. We must also spur the development and deployment of low carbon vehicle and fuel technologies with supportive policies to harness the power of American ingenuity and market forces.

Confronting climate change is a top priority for the Obama Administration. The U.S. Department of Transportation is committed to action that will reduce greenhouse gas emissions, diminish our dependence on foreign oil, create clean energy jobs, build livable communities, and protect us all from dangerous climate change.

The DOT is already taking action through the Department's livability initiative and the Sustainable Communities Partnership with EPA and HUD. The initiative supports low carbon transportation options such as public transportation, walking, and biking; promotes development of housing in close proximity to transit; and promotes mixed-use development that enables residents to easily access goods and services. These actions improve quality of life, lower household transportation budgets, and as shown by this study, reduce greenhouse gas emissions. The Department's high speed rail initiative will also provide a low carbon travel alternative.

Further, in April 2010, the Department and EPA announced national greenhouse gas and fuel economy program for cars and light-duty trucks. Analysis indicates cumulative industry greenhouse gas reductions of approximately 900 million metric tons CO2e and fuel savings of approximately 1.8 billion barrels of oil. The DOT is also implementing new statutory authority to issue fuel economy standards for medium- and heavy-duty trucks. In aviation, DOT has put energy and environmental concerns at the heart of the effort to modernize the U.S. air traffic system, called NextGen. Likewise, the Maritime Administration is focused on the potential of new technologies to reduce the harmful emissions from marine diesel engines through cooperative efforts with the EPA and the maritime industry.

Yet there is more to be done. As indicated in this report, a full range of strategies can be brought to bear to reduce transportation's greenhouse gas emissions: improving fuel efficiency; expanding the use of low-carbon fuels; improving the efficiency of the transportation system; and reducing the volume of travel that relies on carbon-based fuels. These strategies can be implemented through a range of policy options—an economy-wide carbon price, efficiency standards, market-based incentives, transportation planning and investment, and research and development. The DOT looks forward to working with Congress on transportation policy that reduces greenhouse gas emissions, provides for economic vitality, and enhances our quality of life.

www.ingramcontent.com/pod-product-compliance
Lightning Source LLC
Chambersburg PA
CBHW080641180526
45168CB00008B/3250